I0062888

COOL CAREERS WITHOUT COLLEGE FOR
MATH AND
SCIENCE
WIZARDS

COOL CAREERS WITHOUT COLLEGE FOR
MATH AND
SCIENCE
WIZARDS

BETTY BURNETT, Ph.D.

The Rosen Publishing Group, Inc.

New York

Published in 2002 by The Rosen Publishing Group, Inc.
29 East 21st Street, New York, NY 10010

Library of Congress Cataloging-in-Publication Data

Burnett, Betty, 1940–
Cool careers without college for math and science wizards / Betty Burnett.
p. cm. — (Cool careers without college)
Includes bibliographical references and index.
ISBN 978-1-4358-8812-8
1. Mathematics—Vocational guidance—Juvenile literature. 2. Science—Vocational guidance—Juvenile literature. [1. Mathematics—Vocational guidance. 2. Science—Vocational guidance. 3. Mathematics—Vocational guidance. 4. Vocational guidance.] I. Title. II. Series.
QA10.5 .B85 2002
602.3—dc21

2001004475

Manufactured in the United States of America

CONTENTS

INTRODUCTION

Are you the scientific type? Do you like to figure things out? Are you interested in the world around you? Not all scientists have a college degree or their own private laboratory—nor do they need them.

The scientific personality is curious and likes to think through problems. Does that sound like you? Is "What if?" one of your favorite questions? You may like science fiction movies but you may be more interested in how the special effects are done than in the plot. You might find the idea of alien encounters exciting, but you're skeptical about reports of abductions.

Scientific types have a "show me" attitude. They want proof. They want logic. They want to work with gizmos and gadgets, numbers and formulas. Their "people skills" may not be as high on their list as their skill at tinkering. Sometimes they prefer to work alone, but they also like to work with others on a research team.

Scientists use the scientific method to solve problems. First they ask a question: Why doesn't this computer work? They come up with a tentative answer: It's not plugged in. They test their theory by first seeing if the computer is plugged in. If it's not, they plug it in and flip the switch. If the computer starts up, they were right. If not, they were wrong and they start over with another theory to test.

All the sciences are based on the big four: mathematics, physics, chemistry, and biology. A few of the divisions are:

- Anthropology—the study of people and culture
- Astronomy—the study of the stars, the planets, and galaxies
- Atmospheric science—the study of the weather
- Biochemistry—the study of the chemistry of living things
- Botany—the study of plants
- Geography—the study of landforms and maps

- Geology—the study of the structure of the earth (rocks and minerals)
- Metallurgy—the study of metals
- Oceanography—the study of the oceans
- Optics—the study of light
- Psychology—the study of animal and human behavior
- Zoology—the study of animals

Science done for the sake of creating a useful product or process is applied science or technology. Every science has a technology. Botany is the science of plants; horticulture is its technology. Geology is the science of the structure of the earth; mining is one of its technologies, gemology is another.

Technology uses tools and techniques for useful purposes—to send a voice from Seattle to Chicago, or to keep ice cream frozen. For years, technologists and technicians have been giving us the fruits of their labor: smaller radios; larger TVs; faster ways to cook food; better food to cook fast; lightweight jackets that are warmer than heavy ones; cars that know where they're going; and assembly-line robots that never get tired or bored.

Jobs in technology will continue to grow in the years ahead, but they will change faster than they've ever changed before. One of the most important skills for a technician to have is the ability to adjust to change. Few of today's jobs will be the same in ten years. If you want a job in technology, be prepared to keep on learning.

All of the sciences employ technologists and technicians. A technologist has a broad knowledge of the field and can perform many functions. A technician is a person skilled in the use of a specific technique and specific tools. Today's technologists frequently have a college degree, but most technician positions do not require four years of college. They do require training, however, whether from a technical institute, community college, through distance learning, or by way of on-the-job training or apprenticeship.

All technicians must be:

- Computer literate
- Math savvy
- Quick at learning how to use equipment
- Able to communicate clearly with both the written and spoken word

In this book we will look at some of the jobs open to technicians today. There are hundreds more, and new ones are created every year. The easiest way to find a job in the technological field is to talk to people (network); attend conferences and meetings of local science and technology organizations (a robotics club, for instance); go to job fairs; and search out professionals in your field of interest and ask them how they got started.

1

AGRICULTURE TECHNICIAN

You don't have to live on a farm to have an interest in agriculture. Today, agriculture means business—agribusiness—and new advances in the field affect every corner of the globe. Feeding the world's ever-expanding population is this century's most serious concern. Technology will play a major role in the agriculture of the

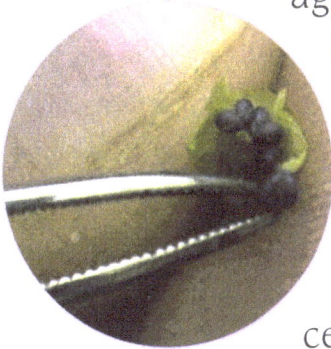

future. Agribusiness involves every aspect of food—growing, processing, preserving, transporting, and distributing it. There are jobs for technicians in all of these areas.

Agriculture technicians work with foods, fibers, and animals. They may work for a pet food company trying to find a dog food that will eliminate "dog breath"; a textile company trying to discover a new synthetic all-weather fabric; a food company hoping to make the ultimate breakfast cereal, which is full of vitamins and better tasting than the other brands; a university lab developing a new strain of wheat; or with any of the hundreds of other companies and agencies in the industry.

If they work in a lab, agriculture technicians assist scientists in running experiments and keeping track of the results. As with all lab work, it is important to be deliberate, accurate, and impartial. Other agriculture technicians work in the field with scientists who are trying to improve the stock of cattle, pigs, chickens, and other domestic animals. Through choosing which animals can mate, they use selective breeding to strengthen a certain strain of animal or to make sure that certain characteristics will show up, such as thicker wool in sheep. Although they work with animals, agriculture technicians must maintain a scientific detachment and never forget that animals are a source of food and will be used in experiments.

Texas A&M University researcher Kevin Crosby demonstrates the technique of cross-pollination he used to develop a variety of habanero pepper that is very mild.

Agricultural technicians who work in the field may do crop surveys to find out which seeds get the best results, or follow-up studies on techniques like hydroponic gardening (growing plants in a nutrient liquid without soil). They test herbicides and pesticides for both their usefulness and their toxicity, the tendency to overkill. They are frequently asked to help solve problems, to determine the cause of soil erosion on someone's farm, for instance, or to test the quality of groundwater. Wherever they work, they must keep accurate

Biotechnology

Biotechnology is a hot topic today—and a very controversial one. It is the technology of combining or manipulating living organisms for a purpose. People have been doing this for hundreds of years, in making beer, wine, cheese, or yogurt. Today, biotechnology has gone beyond the simple combination of organisms and is used to produce biodegradable plastics and new kinds of genetically engineered food that may help solve the problem of world hunger.

Genetic engineering is biotechnology's most attention-getting tool. Some geneticists (scientists who study heredity) have focused on cloning (producing offspring from a single cell instead of the merging of two germ cells). Others are finding ways to change genes so that certain diseases, such as plant blights, will not take hold. Biotechnology will continue to be in the news in the years ahead.

records of what they do and what they observe, reporting back to their employers in careful detail.

A good job for an agricultural technician who likes to travel and meet people is as a field representative for a

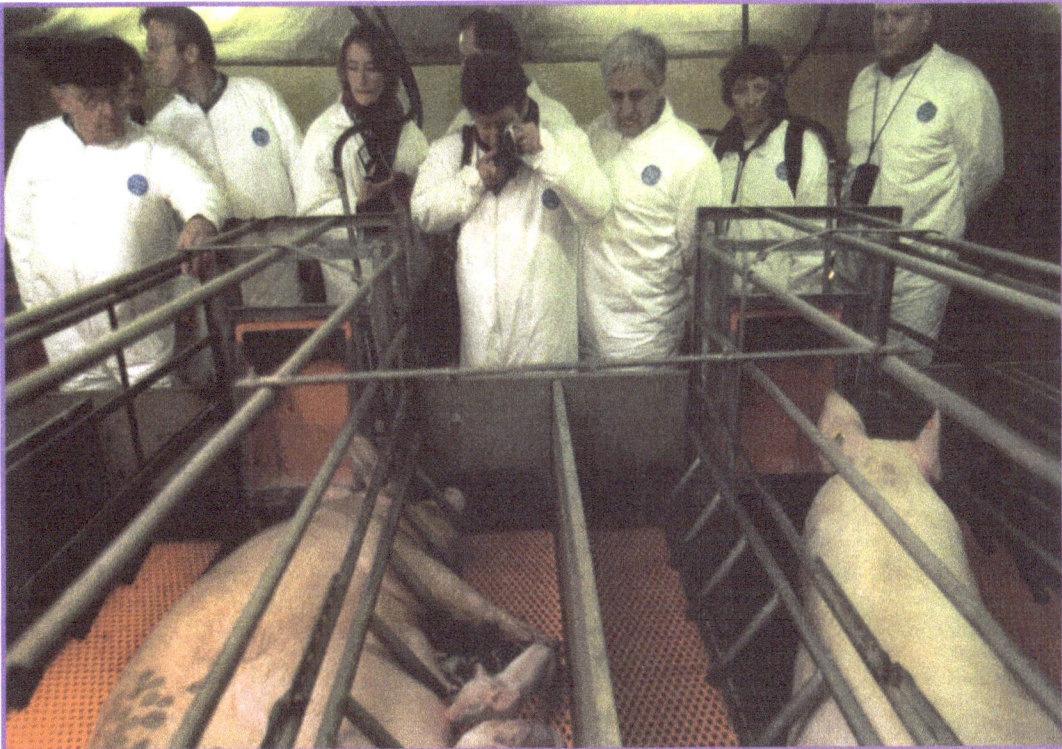

A group of agriculture and food specialists visits a hog farm in Jacksonville, Florida, to discuss developments in biotechnology and the potential effect of genetically modified organisms on human health.

company making farm implements or fertilizers. These technicians demonstrate new techniques and products.

In Canada, where there is an emphasis on sustainable (planet-friendly) agriculture, agricultural technicians are involved in restoring natural habitats by seeding grasses, planting trees as windbreaks, and revitalizing old soils.

Education and Training

Someone with an interest in agriculture should take biology, chemistry, and mathematics classes in high school. Some companies require no formal schooling beyond high school and offer to train new recruits on the job. Others prefer a year or two of further training in the life sciences, chemistry, or animal husbandry. Further education can be obtained online or through conferences and seminars pertaining to the field.

Salary

The salary for an agricultural technician depends on the job and employer. Those who go to work for a county extension agency can expect to start at about $14,000 per year. New employees with a strong science background who work for a large agribusiness could start at an annual salary as high as $30,000.

Outlook

The outlook is very bright for agribusiness, as the pressure increases to find new sources of food for both humans and animals and new ways to produce and distribute food. Jobs for agricultural technicians will continue to be available at large food-processing companies, with research centers, and with federal, state, and county governments.

FOR MORE INFORMATION

ASSOCIATIONS

American Agricultural Economics Association
415 S. Duff Avenue, Suite C
Ames, IA 50010
(515) 233-3202
e-mail: info@aaea.org
Web site: http://www.aaea.org
The association offers a free brochure, "Careers for the Future."

Council for Agricultural Science and Technology (CAST)
4420 W. Lincoln Way
Ames, IA 50014-3447
(515) 292-2125
Web site: http://www.cast-science.org
The Web site for this organization provides scientific information about advances in agriculture and lists workshops and conferences.

Institute of Food Technologists
221 N. LaSalle Street, Suite 300
Chicago, IL 60601-1291
(312) 782-8424
Web site: http://www.ift.org
IFT offers a booklet "Finding Your First Job in Food Science" for $3 (U.S.) postpaid. It is available from the Employee Services Department at the above address. Their brochure "World's Largest Industry" is free.

U.S. Department of Agriculture
1400 Independence Avenue SW
Washington, DC 20250
(202) 720-2791
Web site: http://www.usda.gov
This government organization serves all Americans who benefit from the country's land (meaning, everyone). It is responsible for livestock, produce, water, national forests, conservation efforts, rural areas, research and technology, and much more.

WEB SITES

Agriculture.com
http://www.agriculture.com
The ultimate site for anything relating to agriculture.

Canadian Agriculture Library
http://www.agr.ca
The official Web site for agriculture and agri-food business in Canada offers a good overview of Canadian agriculture and has links to other sites of interest to the career seeker.

Food and Agriculture Organization of the United Nations
e-mail: FAO-HQ@fao.org
Web site: http://www.fao.org
The FAO was organized to raise levels of nutrition and standards of living, to improve agricultural productivity, and to better the condition of rural populations. Today, FAO is the largest specialized agency in the United Nations system and the lead agency for agriculture, forestry, fisheries, and rural development.

4-H
http://www.4-H.org
The Web site for this agriculture-oriented youth organization offers links to many areas of interest, including entomology, animal science, and gardening.

BOOKS

Collins, Donald N., Adrian A. Paradis, and William C. White. *Opportunities in Farming and Agriculture Careers.* Lincolnwood, IL: VGM Career Horizons, 1996.
This book will give you a broad idea about the many different career opportunities for those with a love of the land.

Manning, Richard. *Food Frontiers: The Next Green Revolution.* New York: North Point Press, 2000.
A good overview of the future challenges in agriculture, written by a journalist who takes a global approach to the subject.

Yehling, Carol, ed. *Careers in Focus: Agriculture.* Chicago: Ferguson Publishing, 2001.
This book describes what a life spent working in the field of agriculture includes.

PERIODICALS

Agricultural Research
5601 Sunnyside Avenue
Beltsville, MD 20705-5130
(301) 504-1651
Web site: http://www.ars.usda.gov/is/AR
This monthly trade magazine is published by the Agricultural Research Service.

2

EARTH SCIENCE TECHNICIAN

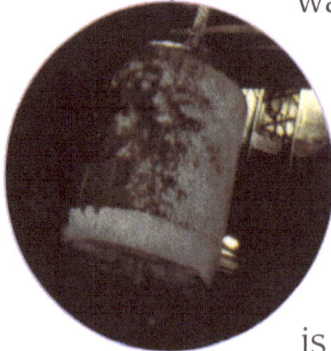

If you ever collected rocks or stopped to admire the rock strata shown in a highway cut, you have an interest in geology or earth science. The earth sciences are concerned with the planet's physical resources—its rocks, minerals, and soil. The technology associated with earth science is mining, or, in the case of oil and natural gas, drilling.

Earth scientists and technicians frequently travel to exotic locations in search of minerals and oil or to study earth forms. "Exotic" does not mean comfortable or pleasant. Minerals are usually found in rugged terrain, miles from cities, and explorers have to work in primitive conditions under extremes of temperature.

Earth science technicians can expect to wear boots and hardhats rather than suits and designer shoes. Their work is strenuous and demanding. Because of this, they often work on contract for short periods of time and receive excellent pay.

Petroleum technicians specialize in oil and gas. They investigate and measure the geologic conditions in working oil or gas wells. To do this, they lower instruments into wells and carefully record their findings. By analyzing the mud from wells, they can tell how much longer the well will continue to produce.

In exploring for new wells (prospecting), technicians collect geological samples and examine them to evaluate the petroleum and mineral content. Some petroleum technicians are scouts. They collect information about existing oil and gas well-drilling operations in the area and about the contracts that leaseholders have. They may suggest that an oil company purchase the lease and develop the site. Petroleum technicians also work on the site as the oil or gas is pumped to a refinery, or they may work in a refinery, where the oil is

Miners construct a 2,000-foot-deep shaft to be used to distill oil at the Occidental Petroleum and Tenneco oil shale mine in Rifle, Colorado.

turned into a commercial product. Refineries have been described as hostile places, with huge, hot pipes and air thick with the smell of oil and gasoline. Working in such an environment requires physical strength and stamina.

Metallurgy technicians, also called materials science technicians, work with mining engineers and frequently travel the globe to find mineral samples for analysis. Usually they are looking for precious metals and valuable elements, such as uranium or gold. They, too, may work on contract for a year at a far-distant place, then come home for a period of time before heading out to the field again. Having the ability to speak several languages is an asset.

Some earth science technicians turn to environmental protection and work on mining reclamation projects, or work at recovering metals from waste streams. Other earth science technicians work with metallurgy engineers. They work in labs or production facilities that make metal products, such as wire, cables, I-beams, or sheets of aluminum. In materials testing labs, the technicians work with metals—primarily iron, steel, aluminum, or copper—to test combinations (alloys) for strength and durability.

Education and Training

Earth science technicians are frequently trained on the job, but they should bring a background in geography or geology, chemistry, physics, and mathematics with them. It can be helpful to take a few courses in these subjects at a community college or university. Technical schools in oil-rich regions teach the basics of petroleum technology.

Salary

Salaries vary widely in this field. There is usually special compensation for the willingness to travel and work

Gemologists

A great way to combine a love of rocks and a love of beauty is to become a gemologist, a person who makes jewelry using precious stones—diamonds, rubies, emeralds, amethysts, and dozens of others. Gemologists design jewelry, cut gems, set and polish stones, and repair broken items. The basic principles of gemology and the use of jewelers' tools can be taught in about six months, and Internet courses are available. It is recommended that anyone interested in jewelry making learn CAD (computer-aided design).

About one-third of gemologists today are self-employed; others work for a jeweler or a manufacturing plant. The median annual income in this field is just under $25,000 per year.

A gemologist looks at a diamond through a small magnifier.

Many petroleum technicians work in refineries, such as the New Jersey oil refinery pictured here.

under difficult conditions. Those who work on-site, whether in oil, gas, or metallurgy, can expect to start at an annual salary of about $18,000 and top out at around $30,000 per year.

Outlook

Earth science technology is a growing field, but it is dependent on the state of the global economy and international politics. Opportunities for those who want to explore the earth are best for those willing to contract with a multinational company, but they have to be prepared for the risks involved, not only from political hotspots in places like the Mideast, but also from unfriendly bacteria and harsh climates.

FOR MORE INFORMATION

ASSOCIATIONS

American Petroleum Institute
1220 L Street NW
Washington, DC 20005
(202) 682-8000
Web site: http://www.api.org
This site is presented like a news magazine, full of information on the industry and links to important issues, such as safety and environmental controls.

Canadian Institute of Mining, Metallurgy, and Petroleum
3400 de Maisonneuve Boulevard W, Suite 1210
Montreal, PQ H3Z 3B8
Canada
(514) 939-2710
e-mail: cim@cim.org
Web site: http://www.cim.org
The excellent Web site for this organization has career information and dozens of links to Canadian institutions. There is also an online magazine.

Gemological Institute of America
5345 Armada Drive
Carlsbad, CA 92008
(800) 421-7250
Web site: http://www.gia.edu
The Web site offers information on job opportunities and training programs for jewelers.

Petroleum Technology Alliance of Canada (PTAC)
Suite 750 Hanover Place
101 6th Avenue SW
Calgary, AL T2P 3P4
Canada
(403) 218-7701
Web site: http://www.ptac.org
The mission statement and purpose of PTAC is "facilitating innovation, technology transfer and research and development in the upstream oil and gas industry."

Society for Mining, Metallurgy, and Exploration (SME)
8307 Shaffer Parkway
Littleton, CO 80127-4102
(303) 973-9550
e-mail: sme@smenet.org.
Web site: http://www.smenet.org
This international organization for earth scientists offers a free publication, "Careers in the Minerals Industry," which may be ordered from its Web site.

WEB SITES

Center for International Earth Science Information Network (CIESIN)
http://www.ciesin.org
CIESIN specializes in global and regional network development and science data management. The site has "metaresources" and a "metadatabase."

The Earth Sciences Portal
http://sdcd.gsfc.nasa.gov
This site, maintained by NASA, is a gateway to general information about the earth.

Earthworks
http://earthworks-jobs.com/
This is a great site for anyone interested in the earth sciences, including oceanography and astronomy. It contains career information and a list of job openings, including contract jobs, plus a lot of links. Earthworks also maintains a book and video store.

BOOKS

Dixon, Dougal. *The Practical Geologist.* New York: Simon and Schuster, 1992.
A clearly written, extensively illustrated guide to everyday geology, this book is highly recommended.

Smith, Peter J. *The Earth.* New York: MacMillan, 1986.
A beautifully illustrated geography-geology book that introduces its subject well.

3

WASTEWATER TECHNICIAN

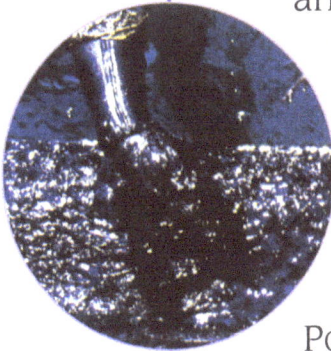

Cleaning up the environment and keeping it clean will take a whole army of scientists and techies in the years ahead. Environmental science covers much more than cleaning up the rivers so fish can live in them or picking up roadside trash on a Saturday morning. Pollution threatens our water, air, and land. Having a part in this massive operation could be very satisfying.

Environmental technicians perform lab and field tests to monitor air and water quality and determine if contaminants are present. They then try to find the sources of pollution. They may also be involved in stopping it (pollution abatement), controlling it, or remedying it. If there is an offshore oil spill, for instance, environmental technicians are sent to the scene to help with the cleanup.

One of the best opportunities for those interested in the environment today is in the field of wastewater treatment. An expanding population with increasing density in urban areas plus more complex manufacturing processes means that our water quality is at risk. Hundreds of chemicals, as well as sewage and discarded food, go into America's wastewater every day.

Wastewater technicians typically work for municipal water treatment plants, but some also work for manufacturers who have orders to meet federal and state guidelines for their effluent (wastewater discharge). Water is pumped from wells, rivers, and streams to water treatment plants, where it is treated and distributed to customers. Wastewater travels to treatment plants via sewers. After water is treated (cleaned) it is returned to streams, rivers, and oceans, or reused for irrigation and landscaping.

An environmental technician participates in an oil spill cleanup.

At water treatment plants, technicians monitor meters and gauges to make sure treatment processes are working properly. They oversee the use of chlorine and other chemicals to purify water. They test the water for the presence of bacteria and parasites. If they find them, they must notify health officials and step up water purification procedures. The Safe Drinking Water Act of 1974 established standards for drinking water; it was upgraded in 1996.

To keep water flowing to homes and businesses, technicians must adjust intake and outflow mechanisms as pressures increase or decrease. At commercial time during the Super Bowl, for instance, there is a tremendous upsurge in water use all across the nation. Technicians must monitor the weather, too. Heavy thunderstorms or long periods of heat and drought call for special procedures in water management.

Some state water-pollution control agencies hire technicians to inspect wastewater treatment plants throughout the state. This requires travel and knowledge of the state's geography, particularly its water resources.

At manufacturing plants, wastewater technicians work under the supervision of chemists and engineers to take steps to counteract the polluted effluent. Almost 50 percent of our serious pollution comes from industrial sites, and most major companies employ environmental engineers and technicians to work at their plants on pollution prevention and control.

Education and Training

Completion of a one-year certification program in water quality and wastewater treatment technology is recommended. It is also important to have a mechanical aptitude. Other skills can be learned on the job. Operators of wastewater treatment facilities must pass an examination to certify that they are capable of overseeing the process.

Volunteering for the Planet

Hundreds of volunteer organizations monitor water quality across North America, from the Rio Grande on the Texas-Mexico border to the Mackenzie River in northern Canada. Their members are constantly checking up on the health of streams, rivers, lakes, reservoirs, coastal waters, wetlands, and wells; they report their findings to government agencies.

Volunteering for such organizations is a good way to get practical experience in environmental care and to learn the basics of water management. Monitoring water resources has united sport fishers, hunters, hikers, bikers, campers, and environmentalists.

Volunteers first note the condition of the area around the body of water they are monitoring. What is the land used for? Is there any protection from flooding? They list the living creatures in and around the water. Then they measure the physical and chemical characteristics of the water: temperature, clarity or cloudiness, rate of flow, amount of oxygen in the water, and the presence of other chemicals. Is there a strange smell? A peculiar color? Before they leave the scene, volunteers also clean it up.

Technicians of Paraguay's National Technology Institute take samples of water and mud on a ranch in Chaco, Paraguay, in search of toxic waste.

Different levels of certification are available to technicians. Ongoing education is a must, as both the technology and the chemistry of water treatment change rapidly.

Salary

The median salary for wastewater plant operators is just under $30,000 per year. For technicians who start in this field, the pay is about $20,000 per year.

Outlook

Growth in this field will continue to be strong. The number of applicants in this field is generally low, so prospects should be good for qualified applicants. Government regulations and industry standards change almost weekly in this field, so flexibility is an asset. Wastewater technician jobs can be found with disposal companies, in industry, and with federal, state, and local government agencies. Employment in private facilities is expected to grow especially quickly.

FOR MORE INFORMATION

ASSOCIATIONS

American Water Works Association (AWWA)
6666 W. Quincy Avenue
Denver, CO 80235
(303) 794-7711
e-mail: info@awwa.com
Web site: http://www.awwa.com
The American Water Works Association, an organization of wastewater professionals, offers online courses on hydraulics, filtration, applied math, and other pertinent subjects.

Association of Boards of Certification (ABC)
208 Fifth Street
Ames, IA 50010-6259
(515) 232-3623
e-mail: abc@abccert.org
Web site: http://abccert.org
The ABC can provide information on becoming certified as a waste-water technician (United States and Canada).

Women in Technology International (WITI)
6345 Balboa Boulevard, Suite 257
Encino, CA 91316
(800) 334-9484
Web site: http://www.witi.org
This organization offers special encouragement to women who want to succeed in technology.

WEB SITES

Bureau of Apprenticeship and Training
http://bat.doleta.gov
Devoted to information on apprenticeships in the United States, this site is maintained by the U.S. Department of Labor, Bureau of Apprenticeship and Training (BAT).

Environmental Protection Agency (EPA)
http://www.epa.gov/ebtpages/water.html
The Environmental Protection Agency's water site is extensive. There is an online reference library that covers every aspect of water: polluted, purified, below or above ground, fresh or salt.

Human Resources Development Agency of Canada
http://hrdc-drhc.gc.ca
An extensive and easy-to-use Web site on technology jobs, including learning and training resources, career planning tools, and self-assessment and aptitude testing.

Wateronline.com
http://wateronline.com
A great deal of information about water resources in general. The industry newsletter is also available at this site.

BOOKS

Career Information Center. *Engineering, Science, and Technology.* Vol. 6. New York: MacMillan Library Reference USA, 2000.
Basic information about technical jobs, including salary, outlook, and training.

Cosgrove, Holli R., ed. *Exploring Tech Careers.* Chicago: J.G. Ferguson Publishing Co., 1995.
A reference directory with loads of information about many technological careers.

Fasulo, Michael, and Paul Walker. *Careers in the Environment.* Chicago: VGM Career Horizons, 2000.
Contains extensive contact information for professional associations, forecasts, salary information, and much more.

"Look Ahead, Get Ahead."
This excellent booklet about emerging careers in technology is online (available as a CD or booklet) from Canadian Technology Human Resources at http://www.cthrb.ca.

Pielou, E. C. *Fresh Water.* Chicago: University of Chicago Press, 1998.
The author blends technology and nature, physics and chemistry with her own practical observations of the water cycle.

4

HAZARDOUS MATERIALS TECHNICIAN

A rapidly growing field for technicians with an interest in environmental protection is the removal and disposal of hazardous materials. "Hazmats" are anything judged a danger to public health and safety and include lead, asbestos, radioactive material, and a host of deadly chemicals, such as corrosive acids and PCBs. Frequently, hazardous materials technicians are the ones called to clean up a toxic spill when there's been a truck, railroad, or shipping accident.

This is a high-stress, risky job, like fire fighting or being part of a bomb squad. You never know where the danger is coming from, when it will come, or exactly what it will be, but you know it's coming. You have to be prepared for anything, including long stretches when nothing happens, or overtime when too much happens at once. Those who work in the emergency field are called hazwopers (hazardous waste operations emergency response).

Technicians work at the site of the contamination, which means they must be able to travel at a moment's notice and perhaps stay away from home for days. Sometimes contaminated sites are evacuated and cordoned off.

In detoxifying a site, technicians use a variety of tools and equipment, from brooms to fire hoses. They frequently have to wear protective gear such as coveralls, gloves, hardhats, safety glasses, and face shields, to protect themselves from the danger of contamination. Sometimes they carry respirators as well.

The training for hazmat technicians is intensive. There are hundreds of toxic chemicals and each one requires special handling. Biological waste from hospitals and clinics is

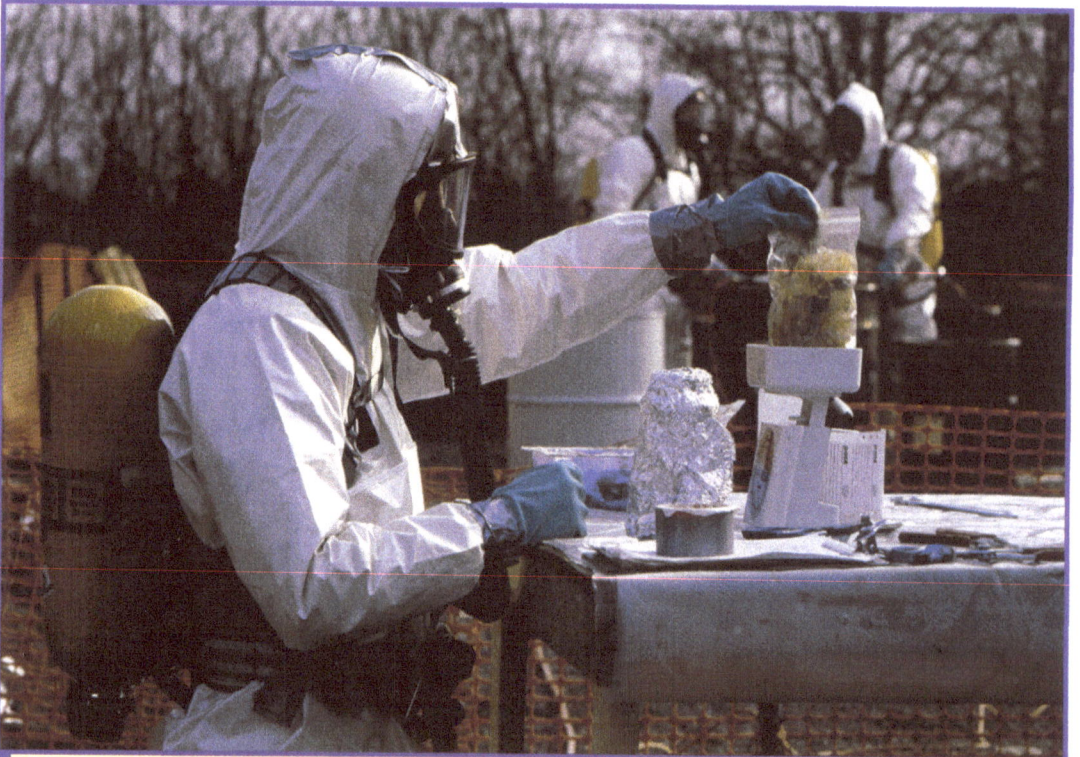

Hazardous materials technicians conduct waste testing and waste removal exercises.

a danger, too. It includes materials contaminated by bacteria and viruses, some of them deadly.

Strict procedures must be followed in the disposal of hazardous wastes, as mandated by law and enforced by the Environmental Protection Agency (EPA). In a lead abatement project, technicians strip lead-based paint from the walls of a home. They put the contaminant in an impregnable container which is then transported to a storage site. Asbestos, which was once used for fireproofing and insulation, too, is considered a hazardous material and must be carefully removed and discarded.

Radioactive material is also a major problem. High-level radioactive waste comes from nuclear reactors and certain industrial sites. If it isn't contained, it threatens the health and safety of everyone for miles around. Nuclear accidents are rare, and reactors have their own safety procedures for containment and cool down. The greater danger is from radioactive material that is being transported.

Low-level radiation is found on hospital uniforms and medical equipment. D&D (decommissioning and decontamination) technicians use radiation survey meters to locate radioactive materials and evaluate how "hot" they are. They wear radiation-sensitive badges to let them know when they're hot and must wash down. To decontaminate a radioactive site, technicians use high-pressure cleaning equipment. TSD (treatment, storage, and disposal) technicians package radioactive waste in shielded containers for disposal at an approved site.

New hazmats will likely come to public attention in the years ahead.

Education and Training

Formal education beyond a high school diploma is not required for most of this work. A background in math is essential, because workers need to perform mathematical conversions and calculations, and figure readings. Federal

Case Study: Dioxin

Dioxin is a hazardous chemical that won't go away. It's produced when municipal wastes, especially plastics, are burned. The EPA lists dioxin in the top 10 percent of substances dangerous to human health. It results in higher rates of cancer, damage to the liver and nervous system, and birth defects.

Carelessly dumped by an industrial waste firm, dioxin was responsible for two disasters in the 1980s, one in Times Beach, Missouri, and the other at Love Canal, New York. Residents were alerted to the danger when high levels of cancers and other diseases appeared in their neighborhoods. The town of Times Beach was completely closed down and no longer exists. Houses were destroyed and the soil dug up, removed, and buried elsewhere.

To prevent such disasters from happening again, the EPA and other agencies monitor the air quality around landfills and incineration sites. When dioxin or other hazardous chemicals are detected, these agencies take steps to end the pollution immediately.

Hazardous materials technicians removing toxic waste

regulations require technicians to obtain a license to work with each type of hazardous material. A license may be obtained after completing a basic training program of about forty hours. For work with radioactive materials, longer training is required. In all cases, technicians must take yearly courses to update their knowledge.

To advance in the field, it is wise to have post-secondary courses in chemistry, environmental toxicology, and civil engineering.

Salary

The pay in hazardous waste removal ranges from $10.50 to $18 an hour for someone without a postsecondary education. As in all technical fields, the more you know, the more you earn.

Outlook

This field is growing, due to increased pressure for a safer and cleaner environment and the increase in toxic and dangerous wastes. Technicians may find employment with local government agencies, with generalized disposal firms, or with specialized hazardous waste removal companies.

FOR MORE INFORMATION

ASSOCIATIONS

Environmental Protection Agency (EPA)
Ariel Rios Building
1200 Pennsylvania Avenue NW
Washington, DC 20460
(202) 260-2090
Web site: http://www.epa.gov
The EPA's mission is "to protect human health and safeguard the natural environment—air, water, and land—upon which life depends."

Laborers-AGC Education and Training Fund
Route 97
Putnam, CT 06260
(800) 974-0800
Web site: http://www.laborers-agc.org
Information on training as a hazardous waste technician is available through this organization.

Women in Technology International (WITI)
6345 Balboa Boulevard
Suite 257
Encino, CA 91316
(800) 334-9484
Web site: http://www.witi.org
This organization offers special encouragement to women who want to succeed in technology.

WEB SITES

EPA-maintained Web sites
http://clu-in.com
The EPA maintains this site for hazardous waste information, which describes the most up-to-date treatments and technologies.

http://www.epa.gov/msw
A more general EPA site, surprisingly upbeat and interesting, is on municipal solid waste (aka trash), including toxic junk.

Human Resources Development Agency of Canada
http://hrdc-drhc.gc.ca
An extensive and easy-to-use Web site on technology jobs, including learning and training resources, career planning tools, and self-assessment and aptitude testing.

National Environmental Trainers
http://www.natlenvtrainers.com
Information on online training programs can be found at the virtual training institute of the National Environmental Trainers.

BOOKS

Career Information Center. *Engineering, Science, and Technology.* Vol. 6. New York: MacMillan Library Reference USA, 2000.
Basic information about technical jobs, including salary, outlook, and training.

Cosgrove, Holli R., ed. *Exploring Tech Careers.* Chicago: J.G. Ferguson Publishing Co., 1995.
A reference directory with loads of information about many technological careers.

Fasulo, Michael, and Paul Walker. *Careers in the Environment.* Chicago: VGM Career Horizons, 2000.
Contains extensive contact information for professional associations, forecasts, salary information, and much more.

Gerrard, Michael. *Whose Backyard, Whose Risk?* Cambridge, MA: MIT Press, 1994.
The subtitle for this book is *Fear and Fairness in Toxic and Nuclear Waste Siting,* and it examines the politics of hazardous waste management.

Mazur, Allan. *A Hazardous Inquiry: The Rashomon Effect at Love Canal.* Cambridge, MA: Harvard University Press, 1998.
A journalist interviewed many of the people involved in the Love Canal disaster and presents several points of view.

VIDEOS

A Civil Action (1999)
Based on a true story, this film, starring John Travolta, brought national attention to the dangers of hazardous waste.

Race to Save the Planet: Do We Really Want to Live This Way? (1990)
Booking Video
Chedd-Angier Production Co.
A sixty-minute documentary that gives an overview of toxic wastes.

5

AEROSPACE (AVIONICS) TECHNICIAN

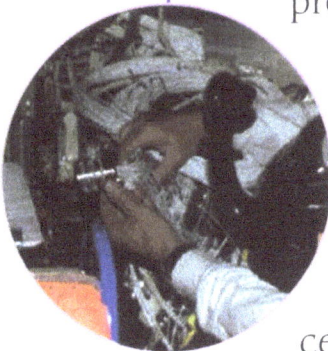

Aerospace technicians assist in the design, development, testing, and production of aircraft and space-craft—rockets, missiles, heli-copters, airplanes, and space vehicles. Because the field is so vast and the equipment so complex, aerospace tech-nicians usually specialize in certain areas. They might spend several years working on one small part needed for a booster and become an expert in that branch of aerospace technology.

Avionics—aviation electronics—is a rapidly growing part of the aerospace industry. Avionics technicians repair and maintain the electronic controls used for aircraft navigation, radio communications, and weather radar systems. Other electronic equipment on board controls flight, engines, and other primary functions.

All aircraft are dependent on instrumentation. Instrument flight regulations (IFR) require airplanes to be equipped with position-finding instruments. For commercial airlines, the key system is the flight management computer, which can display altitude, speed, course, wind conditions, and route information. The instrument landing system (ILS) enables an airplane to navigate through clouds or darkness to an airport's runway. The microwave landing system (MLS) can land the plane automatically if the pilot is unable to, but the pilot always has the option of overriding it manually.

Pilots monitor their cockpit panels closely and immediately report even the smallest glitch. Avionics technicians take over from there. First, they try to recreate the conditions that were present when the malfunction occurred. Then, they must pinpoint the problem and find its source, using complex diagnostic equipment. Diagnostics is also used in preventive maintenance, which is routine today in most airlines.

Avionics installers refit the avionics on a Gulfstream II jet.

Today, it is more likely that aviation electronics will be repaired, rather than replaced. Sometimes the equipment is removed from the plane and taken to a repair shop, but more often avionics technicians are required to work in cockpits in hangars or outdoors on the airport tarmac. They are under pressure to work fast in order to maintain flight schedules. At the same time, they must meet safety standards and not give a thumbs-up until they are absolutely certain the part is in perfect working order. The safety of passengers and crew is the number one priority. Usually

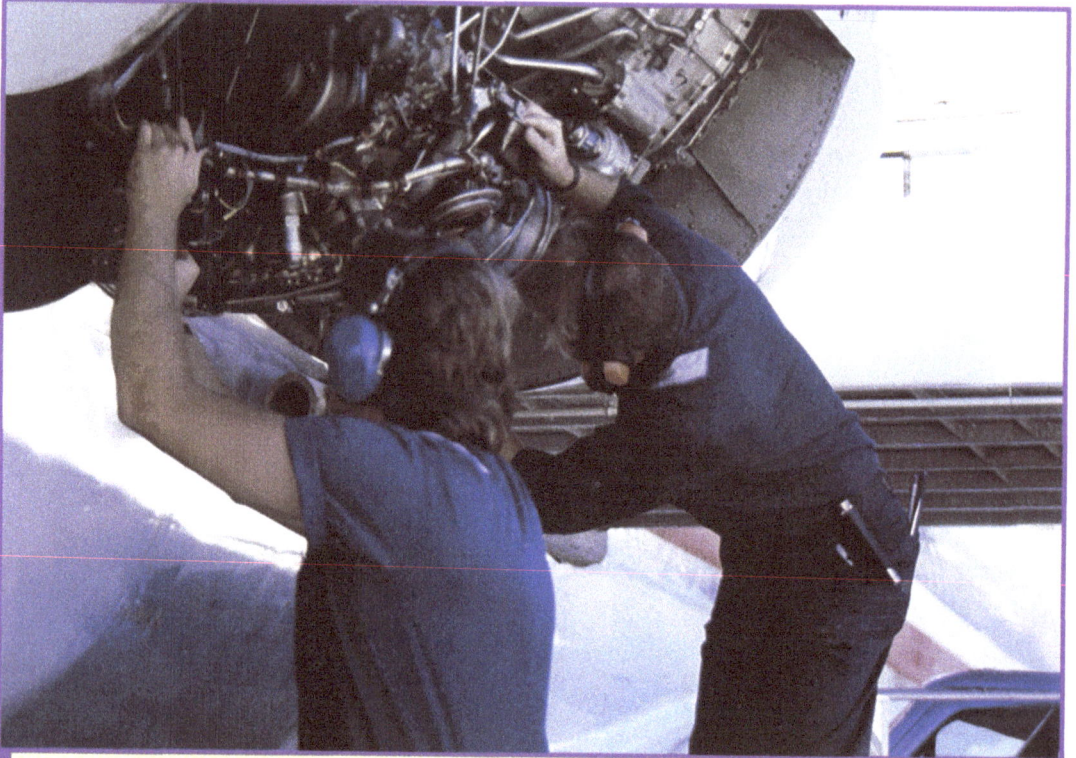

Aircraft engineers make repairs to a plane.

technicians work in teams and share this responsibility. A stress-resistant team player who accepts overtime and weekend work makes a perfect employee.

About two-thirds of avionics technicians work for airlines or airports; the rest work for aircraft assembly plants, repair shops, private aircraft firms, or the federal government.

Education and Training

Avionics technicians usually complete a two-year program in avionics, engineering, or aerospace technology at an

Avionics Alphabet Soup

From National Aeronautics and Space Administration (NASA) to Global Positioning System (GPS), aeronautics technology sprouts initials. Here are some avionics acronyms:

- **EICAS** Engine Indicating and Crew Alerting System. This displays the status of the engine and other features of the aircraft that are important to its safety.
- **FADEC** Full-Authority Digital Engine Control. This controls the fuel for the throttle and the autopilot.
- **HUGS** Head Up Guidance System. This laser-reflective clear glass plate displays information about the aircraft. It is designed so the pilot can look directly out the window while reading the screen.
- **ILS** Instrument Landing System. This aligns the aircraft with the runway on the final approach before landing.
- **IRS** Inertial Reference System. A laser gyroscope that gives the pilot information on the plane's altitude and acceleration.

institution certified by the Federal Aviation Administration (FAA). Courses in physics, chemistry, electronics, computer science, and mechanical drawing are helpful. Many employers offer on-the-job training, and apprenticeships are available. Training by way of the armed forces is also valuable. FAA certification is required for technicians who work at airports and similar installations. Passing a test after at least eighteen months of work experience or completion of a technical program will result in certification. All technicians who work with radios must get a Federal Communications Commission (FCC) license.

Salary

Avionics technicians make between $15 and $22 per hour. Those who work for major airlines generally earn the higher wages and are covered by union agreements, which offer additional benefits.

Outlook

The outlook for this career is highly dependent on the future of the airline industry. The trend toward modernizing and refitting older aircraft with new electronics means that the outlook for avionics technicians should be good in the years ahead. Opportunities will be best with small commuter and regional airlines. Competition will be tough for jobs with the large airlines. Those who keep current with technological advances in electronics composite materials will be most in demand.

FOR MORE INFORMATION

ASSOCIATIONS

Aircraft Electronics Association (AEA)
4217 S. Hocker
Independence, MO 64055
(816) 373-6565
e-mail: info@aea.net
Web site: http://www.aea.net
This excellent site has information about the technical journal *Avionics News*, with sample articles. There is also information on careers in the field.

National Aeronautics and Space Administration (NASA)
Headquarters
300 E Street SW
Washington, DC 20546-0001
(202) 358-0000
Web site: http://www.nasa.gov
The NASA Web site is a source of inspiration and information to anyone interested in aviation or aerospace exploration. NASA is the granddaddy of avionics.

Professional Aviation Maintenance Association (PAMA)
1700 H Street NW, Suite 700
Washington, DC 20006
(202) 730-0260
Web site: http://www.pama.org
This organization is made up of airframe mechanics and avionics technicians. The emphasis is on continuing education in this fast-changing field.

Women in Technology International (WITI)
6345 Balboa Boulevard, Suite 257
Encino, CA 91316
(800) 334-9484
Web site: http://www.witi.org
This organization offers special encouragement to women who want to succeed in technology.

WEB SITES

Avionics.com
http://www.avionics.com
A Web site devoted entirely to avionics.

Avweb.com
http://www.avweb.com
Avweb is an online magazine full of information about the aviation/avionics industry. Employment ads can give you an idea of what jobs are available.

Human Resources Development Agency of Canada
http://hrdc-drhc.gc.ca
An extensive and easy-to-use Web site on technology jobs, including learning and training resources, career planning tools, and self-assessment and aptitude testing.

BOOKS

Career Information Center: *Engineering, Science, and Technology.* Vol. 6. New York: MacMillan Library Reference USA, 2000.
Basic information about technical jobs, including salary, outlook, and training.

Cosgrove, Holli R., ed. *Exploring Tech Careers.* Chicago: J.G. Ferguson Publishing Co., 1995.
A reference directory with loads of information about many technological careers.

Fasulo, Michael, and Paul Walker. *Careers in the Environment.* Chicago: VGM Career Horizons, 2000.
Contains extensive contact information for professional associations, forecasts, salary information, and much more.

Helfrick, Albert. *Principles of Avionics.* Englewood Cliffs, NJ: Prentice-Hall, 2000.
Written by an industry leader, *Principles of Avionics* is used as a textbook in many aviation schools.

Kayton, Myron. *Avionics Navigation Systems.* New York: Wiley, 1997.
This book covers the basics through the recent advances in navigation theory and hardware and software.

Langewiesche, Wolfgang. *Stick and Rudder: An Explanation of the Art of Flying.* Tab Books, 1990.
This is the classic book on the fundamentals of aerodynamics. It is a book pilots and would-be pilots love.

Taylor, Richard L. *Instrument Flying.* New York: McGraw-Hill, 1997.
This book is a detailed and highly interesting description of the technology and advances in avionics.

6

MEDICAL LAB TECHNICIAN

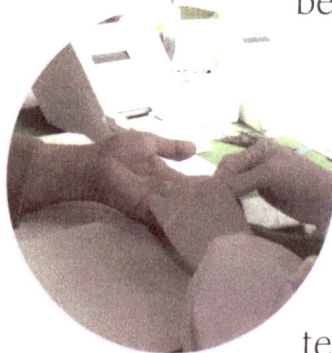

The amazing medical advances made in the last fifty years have come about because scientists have been willing to work long hours in labs, learning, puzzling, and experimenting. Medical technology has played a large role in these advances. Medical technicians and technologists run tests that detect disease. The results they find help in its diagnosis.

Their findings also lead doctors to decide the course of treatment for their patients.

Almost everyone has had at least one medical test, usually a blood test or a drug test. Medical technicians perform these routine lab tests under the supervision of medical personnel. In this field, there is a clear distinction between a technologist and a technician. Medical technologists, with their more extensive training, have more authority to make decisions. Medical technicians are encouraged to follow instructions to the letter and leave decision making to others. They usually work under the supervision of technologists, but sometimes they work for medical doctors.

Medical tests are run on blood, urine, tissue, spinal fluid, cells of various kinds, and whatever else a doctor calls for. Today, much lab equipment is automated and technicians can run several tests at the same time, for blood sugars or hormone levels, for instance. Early detection of disease through such tests has been responsible for lengthening the lifespan of many Americans. The best example of a life-saving test is the PSA test which measures prostate-specific antigen and can catch prostate cancer in its earliest and most treatable stages.

Technologists examine blood and tissue samples under microscopes. Technicians, too, can learn to do tissue (histology) cultures to see if certain organisms are present. Lab

A medical lab technician pricks the finger of North Dakota representative Rachel Disrud during a cholesterol and blood glucose screening at the capitol in Bismarck, North Dakota.

personnel look for bacteria, parasites, evidence of drug use, cancer cells, or anything that looks abnormal. Normal test results do not require that a doctor be called, but if anything abnormal is found, the technologist must flag the specimen for further examination by a doctor.

Medical technicians must be willing to work with specimens that may contain contagious diseases. They are frequently gloved and masked and always use sterile procedures. Autoclaves are used to sterilize lab equipment at high temperatures under steam pressure. Technicians use

Medical Technology and Crime Detection

Medical technology has long played an important role in the detection of criminals. Identification of DNA, hair, and blood can put a suspect at the scene of the crime beyond doubt. A whole specialty of forensic medicine has developed in response to this technology. Now, a tiny subspecialty is emerging: wildlife forensics.

Wildlife forensics, including ballistics and toxicology (the study of poisons), can pinpoint the cause of an animal's death. Animal clues can lead to "whodunit," just as in the case of a human victim. Perpetrators are prosecuted for killing endangered species, for hunting out of season, for poaching, and for transporting animals illegally. This field is expected to grow in the years ahead and will need technicians to do the microscopic analysis.

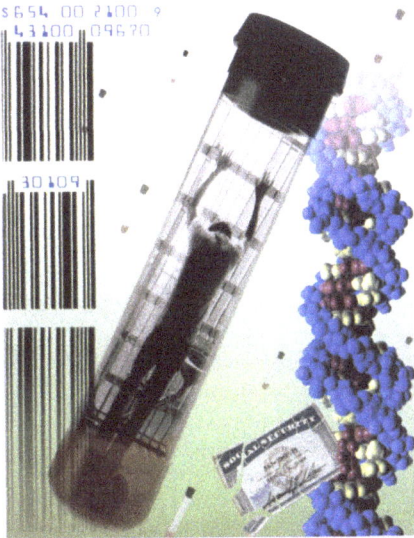

DNA fingerprinting and blood testing can aid investigators in catching criminals.

A technician prepares bone marrow for a transplant.

electron microscopes to do biopsies—microdissections on tiny slices of diseased organs. Other equipment they work with are centrifuges, used to separate blood cells from whole blood; electronic cell counters, which are used to count red blood cells for the diagnosis of anemia; and white blood cells for the diagnosis of infections.

Some technicians, especially those in hospitals, have direct contact with patients and draw blood samples. Others stay in the lab and never see patients. They type blood, determine if it's positive or negative, then match blood types and donors for transfusions. They must keep careful records of all their procedures and the names of patients and their doctors.

Technicians also set up and take down equipment and are responsible for stocking supplies. They should be detail-oriented, careful, and especially neat and clean.

Education and Training

Medical technicians generally have a certificate from a hospital, vocational or technical school, or the armed forces. Certificates are awarded when professional competency meets certain standards, usually after at least one year of training. On-the-job training is very important in clinical work because each institution has its own set of procedures. Some states require licensing or registration for medical technicians who work in hospitals or clinics.

Medical technologists usually have earned a four-year college degree, although it is possible to work up to the position after years of experience as a technician.

Salary

Medical technicians who work in hospital labs can expect to make about $26,000 per year; those who work in private clinics or for government agencies are paid slightly less.

Outlook

Employment for medical lab technicians is expected to grow at an average rate over the next ten years. Because of changes in the medical fields, labs are consolidating and restructuring. Small hospitals often send their specimens to commercial labs, rather than doing lab work in-house. The emergence of more powerful diagnostic tests requires technicians to continue their training while they work. Lab jobs can be found at large hospitals, clinics, blood banks, commercial labs, and research facilities.

FOR MORE INFORMATION

ASSOCIATIONS

American Medical Technologists
710 Higgins Road
Park Ridge, IL 60068
(708) 823-5169
Web site: http://amt1.com
The American association for professionals in allied health care. Membership benefits include access to job postings, scholarships, trade journals, professional conferences, and much more.

American Society for Medical Technologists

7919 Woodmont Avenue, Suite 1301
Bethesda, MD 20814
(301) 657-2768
This association, which also operates on the state level, is the professional organization for medical technologists.

The American Society of Clinical Pathologists (ASCP)

Board of Registry
P.O. Box 12277
Chicago, IL 60612-0277
(800) 621-4142, ext.1345
e-mail: bor@ascp.org
Web site: http://www.ascp.org/bor
ASCP is a not-for-profit medical society organized exclusively for educational, scientific, and charitable purposes. Contact its board of registry for educational requirements and certification requirements.

Commission on Accreditation of Allied Health Education Programs (CAAHEP)

35 E. Wacker Drive, Suite 1970
Chicago, IL 60601
(312) 553-9355
Web site: http://www.caahep.org
CAAHEP accredits programs representing 18 allied health professions recognizing over 1,900 allied health education programs in more than 1,300 institutions.

National Accrediting Agency for Clinical Laboratory Sciences.

8410 W. Bryn Mawr, Suite 670
Chicago, IL 60631
(773) 714-8880
Web site: http://www.naacls.org
This organization is responsible for the accreditation and approval of educational programs in the clinical laboratory sciences and related

health professions through the involvement of expert volunteers and its dedication to public service.

WEB SITES

All the Virology on the WWW
http://www.virology.net
Everything you ever wanted to know about the bugs that make us sick can be found at this easy-to-navigate site full of information of interest to the medical lab technician. There's even a listing of online courses and tutorials on viruses.

Centers for Disease Control (CDC)
http://www.cdc.gov
CDC is America's leading federal agency devoted to the health and welfare of the people. Check out their Web site to learn about what's going on all over the world in the field of disease.

International Association of Medical Laboratory Technologists
http://www.iamlt.org
IAMLT is an international professional organization, which consists of associations of medical laboratory technologists/scientists from forty countries. All members of associations belonging to the IAMLT are automatically members of the IAMLT.

World Health Organization (WHO)
http://www.who.int
This organization's Web site posts current news about health and medicine around the world.

BOOKS

Career Information Center. *Engineering, Science, and Technology.* Vol. 6. New York: Macmillan Library Reference USA, 2000.
This book provides basic information about technical jobs, including salary, outlook, and training.

Cosgrove, Holli R., ed. *Exploring Tech Careers.* Chicago: J.G. Ferguson Publishing Co., 1995.
This is a reference directory with loads of information about many technological careers.

DeSalle, Rob. *Epidemic! The World of Infectious Diseases.* New York: The New Press with the American Museum of Natural History, 1999. This book offers a chilling look at diseases, past, present, and future, with a good discussion of how technology can help detect and stop new epidemics.

Farr, J. Michael. *American's Fastest Growing Jobs*, 6th ed. Indianapolis: JIST Works, 2001
This book contains information on salary, trends, education, and working conditions for more than 100 of the fastest growing jobs in the United States.

Ryan, Frank. *Virus X: Tracking the New Killer Plagues.* Boston: Little Brown, 1997. These stories emphasize medical detection and the important role that lab work will play in stopping the new plagues.

7

ROBOTICS TECHNICIAN

Robotics is the science of remote handling. Robots are modeled on humans, with "arms" and "fingers" that can move in arcs or circles. Some robots can fit tiny parts precisely in place. Others can hold and operate welding torches or paint sprayers. Some are so sensitive they can pick up an egg without breaking it. They can walk, respond to voice commands, and see, using TV cameras for eyes.

The use of robots in assembly lines is changing the way American companies manufacture goods. Robots never call in late or complain about boredom, but they do break down, and that's when a robotics technician is needed.

There are several kinds of robotics technicians, each requiring specific skills. All these technicians need good eyesight, manual dexterity, and the ability to figure things out on their own. This is a good field for people who like to tinker with machinery.

Robot assemblers work with robotics engineers to design and construct industrial robots. They may build a prototype (first of its kind) or tweak an old model. There are hundreds of designs for robots and most of them don't look anything like toy models. Many robot assemblers also build robots as a hobby and stay current with robot electronics as it continues to emerge.

Programmers also work under the direction of a robotics engineer. They help develop the computer programs that will direct the robot to respond to commands and do a specific task. The task may be as simple as picking up a bolt and putting it in place.

Trainers "teach" robots what to do. They use a keyboard called a teach pendant, or controller. Once the desired actions of the robot are chosen, they are keyed in, one by

Some auto manufacturers use robots to build cars.

one. Each movement is stored in the controller's memory. The robot can then repeat the exact movement an infinite number of times. If a programmer comes up with a new set of actions, the robot can be retrained.

Installers set up robots at a manufacturing plant or other site and oversee the operation until they are satisfied that everything is working correctly. Installers usually work for a robot manufacturer and travel where they are assigned. An assignment may last several weeks.

Robot operators, also called precision assemblers, operate robots. They must read and interpret engineering specifications about how the robots should work. These specifications may be in text form, but are often drawings or CAD (computer-aided design) programs. Precision assemblers may work on their feet in old-style noisy assembly lines. The trend today is to work on subassemblies sitting at tables in quiet rooms with small-sized robots.

Precision electrical and electronic equipment assemblers with their robots put together missile control systems, radio equipment, computers, machine-tool numerical controls, radar, sonar, and appliances. Precision electromechanical equipment workers assemble machinery for offices, oil fields, textile plants, printing companies, food processors, and dozens of other industries. They also rebuild engines and turbines.

Education and Training

A high school diploma is generally needed to be hired on an assembly line as a robot operator. Specialized training is recommended for electrical or electronic assembly, and frequently this training is provided by the robot manufacturer. Robot assemblers, programmers, and trainers would benefit from technical training in electronics and computer design. Courses in robotics are offered at some technical schools, and some apprenticeships are available. This field is changing very fast and training requirements will change, too.

Robot History

While most people think of robots as space-age creations, ancient Egyptians made mechanical figures 2,500 years ago. In the eighteenth and nineteenth century, automatons—mechanical devices built in the shapes of people—could move on their own by means of springs. Later these automatons were motorized or battery operated, but they were considered merely amusing toys.

In the mid-1950s, George Devol developed Unimate, an industrial robot that began work on an automobile assembly line. This pioneer robot ran for 100,000 hours and is now retired and on display in the Smithsonian Institute in Washington, D.C.

Little was done in industrial robotics until the development of miniaturization and microchips in the 1990s, which made the field possible.

Outlook

With the success of robots in the auto and airframe industries, many other kinds of assembly lines are converting to allow robots to take over tedious jobs from humans. The projected growth in robotics is 5 percent per year over the next ten years, but that is probably conservative. Both

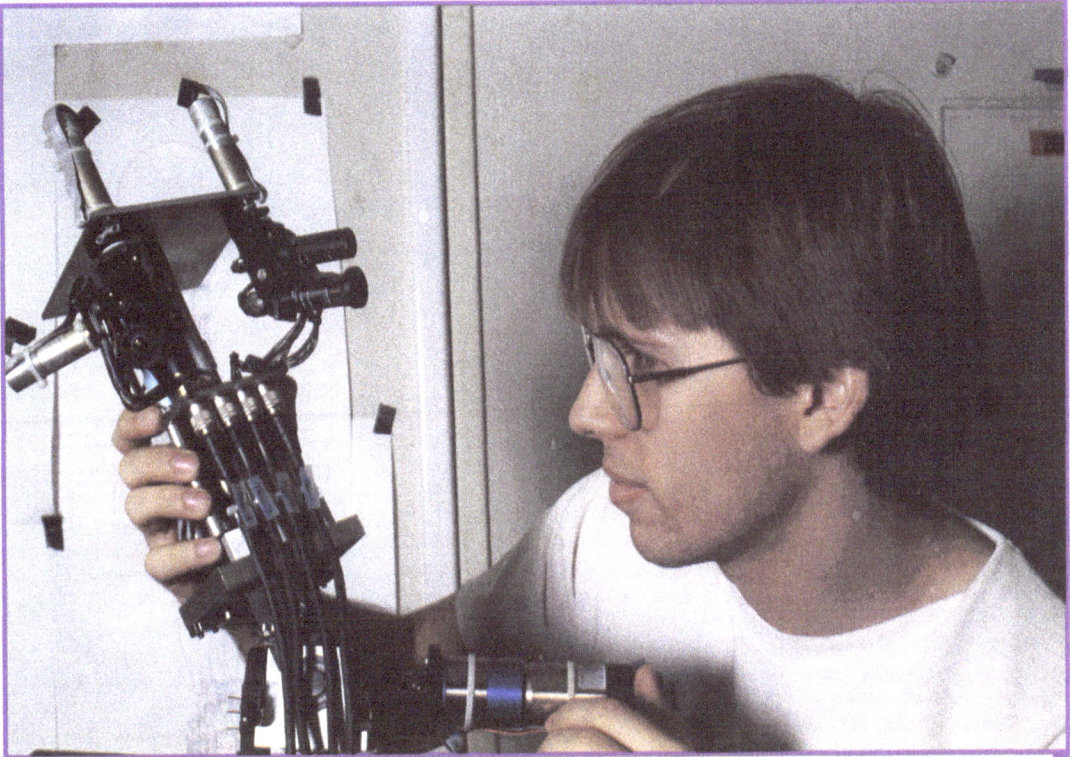

A student equips a robot with cameras to give it vision recognition capabilities.

robot manufacturers and manufacturing firms using robots hire technicians.

Salary

Robotics technicians are paid an hourly wage. Depending on the industry and geographic location, rates range from $6 to $17.50 an hour for entry-level positions. An experienced technician can expect to earn as much as $25 an hour, although $15 an hour is more standard. Since this field is in a major growth phase, it is hard to predict what wages will be in five or ten years.

FOR MORE INFORMATION

ASSOCIATIONS

Robotics Industries Association
900 Victors Way
P.O. Box 3724
Ann Arbor, MI 48016
(734) 994-6088
Web site: http://robotics.org/
This association is all-business. Their extensive Web site offers buyers' guides, educational resources, chats, a career center, and links to everything else.

Robotics International of the Society of Manufacturing Engineers (RI/SME)
P.O. Box 930
Dearborn, MI 48121-0930
(313) 271-1500
Web site: http://www.sme.org
This is a professional organization for those who work with industrial robots. The site is full of technical data and tech trends. There is also information on trade shows and education, a bookstore and a video store. The newsletter of the association is *Robotics Today*.

WEB SITES

ars robotica
http://www.arsrobotica.com
An entire Web site devoted to the robotics industry. Check here for news, editorials, features, and interviews.

Robots.net
http://robots.net/
This site presents the latest information on personal and industrial robotics, although the emphasis is on personal robots. There are chat rooms and news of robot competitions.

Techgeek.com
http://techgeek.com/
A community Web site for robotics professionals and enthusiasts.

BOOKS

Menzel, Peter. *Robo Sapiens: Evolution of a New Species.* Cambridge, MA: MIT Press, 2000.
This is a popular science account of what's going on in robotics around the world, with dozens of photographs. A great read for "bots" lovers.

Wise, Edwin. *Applied Robotics.* Indianapolis, IN: Prompt Publications, 1999.
This is a manual for those who would like to build robots. It has information on the mechanics, sensory systems, electronic controls, and software. The accompanying CD-ROM contains source codes and Atmel development tools.

PERIODICALS

Robot Science & Technology
3875 Taylor Road, Suite B
Loomis, CA 95650
(888) 510-7728
e-mail: service@robotmag.com
Web site: http://www.robotmag.com
This magazine prints in-depth reports about robotics in the work-place, at home, at school, and in sports.

Robotics World
Douglas Publications
2807 N. Parham Road, Suite 200
Richmond, VA 23294
(804) 762-9600
e-mail: info@douglaspublications.com
Web site: http://www.roboticsworld.com
This trade journal features how-to articles, news on industry trends, and case studies of robotics applications.

VIDEOS

Robots Alive! Scientific American Frontiers (1997)
Scientific American.
This video, narrated by Alan Alda, was first aired on PBS on the popular *Scientific American* program. These videos can be ordered from the PBS Web site, www.pbs.org.

8

COST ESTIMATOR

Number crunchers—people who like to work with numbers and mathematics—can aim for a variety of jobs under the general heading of mathematics technician or mathematics support personnel. For such jobs, technicians use math formulas, principles, and methods, as well as simple arithmetic.

All jobs in mathematics require the technician to gather data, analyze it, and apply it to a specific problem. The technician might prepare flow charts, graphs, statistics, profit-and-loss statements, or sales reports. One of the best-paid positions for math lovers is that of cost estimator.

Cost estimators compile data on all the factors that influence the cost of a construction, remodeling, retooling or production job, both materials and labor. The figures they come up with help determine whether or not the job will be profitable. Most cost estimators work for the construction industry or for companies that have major contracts with the government. The estimator-technician, along with the cost engineer, gathers information on the possible prices for utilities, insurance, tools, software development, and a hundred other things that make up the total cost of the project. Executives then analyze the data and decide whether or not to proceed with the project.

In the case of a construction project, the first step is a visit to the site where the new building will be constructed. Are electricity and water easily available? Will there be special problems in digging the basement? Is the ground spongy or is there a rock layer in the way that will require blasting?

Next, the cost-estimating team examines the architect's drawings. What kinds of material are to be used? A granite

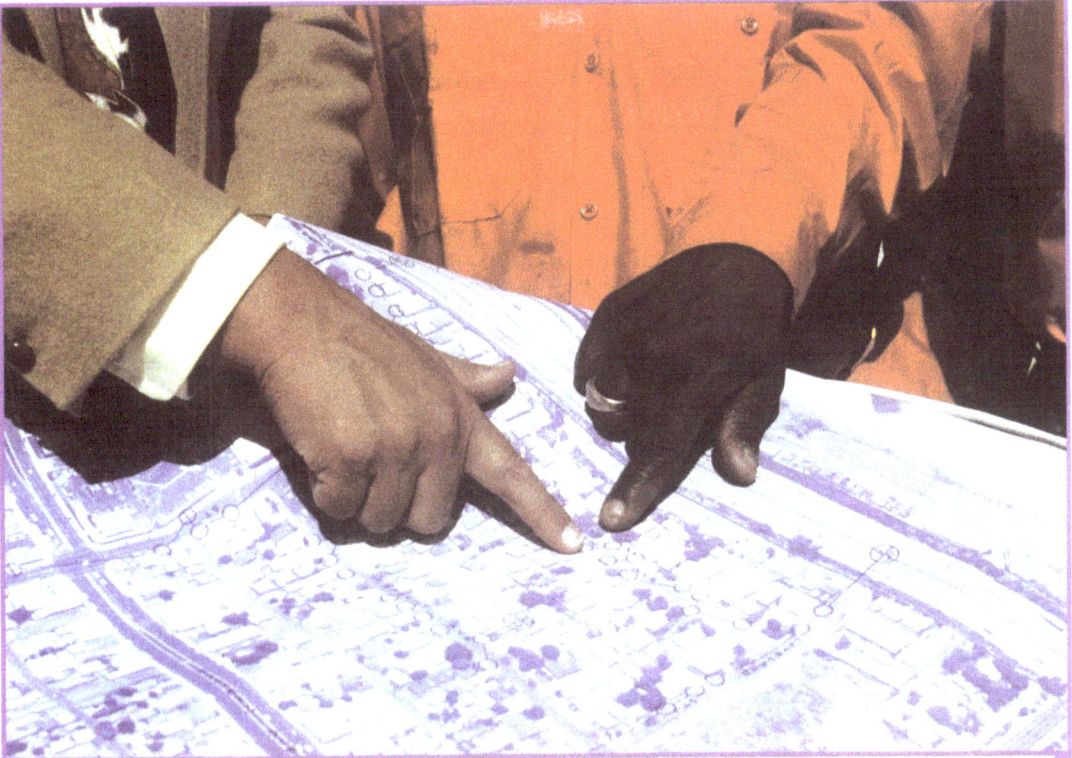

Developers review a blueprint for a construction site.

face will require more work and cost than aluminum siding. How many windows and doors? What kind of roof?

As they sort through the hundreds of details, plugging them into the computer, cost estimators come up with a schedule for completion of the project, factoring in weather delays, of course. They also determine the number of workers needed, both general construction and specialized tradespeople. Using this information, they can calculate the standard labor hours necessary for the job. Standard labor hours are then converted to dollar values. Still to come are the costs of materials (allowing for inflation and substitution).

Attention to detail and a willingness to recheck work are important to cost estimating. So, too, is the ability to handle stress. Everyone wants a job done on time and under budget. That rarely happens.

Education and Training

Experience in the building trades is still the most important asset to getting a position as a cost estimator, but more companies are looking for candidates who have taken construction, engineering, or architecture courses. It is especially important to be able to read blueprints and construction specifications.

All people working with numbers must have strong computer skills and be able to learn software programs quickly. Whether or not they have formal training beyond high school, they need to be able to use algebra, geometry, and trigonometry. In some cases, calculus is also advised.

Salary

Experienced cost estimators earn between $30,000 and $75,000 per year, with the average being around $40,000 per year. It takes many years to work up to this level. An

A Is for Arithmetic

Here is a list of some other jobs for number crunchers.

Actuaries compile statistics about life expectancy and then calculate insurance premiums. They also figure the risks for car, theft, fire, and medical insurance based on hundreds of factors, and compute rates for premiums.

Adjusters investigate billing errors and adjust accounts. They work for retail establishments, such as department stores and credit card companies. Sometimes they are required to explain to customers how the errors have been corrected.

Appraisers give a value to something in dollars. It may be real estate, antiques, or collectibles, such as metal lunchboxes from the 1950s or vintage clothing. Appraisers are frequently self-taught and self-employed. Sometimes, as in the case of real estate, they must pass a test and earn a license. Appraisers who work for insurance companies determine the cost of repairs or damages.

Auditing clerks verify records of transactions posted by other workers. They check figures and documents for mathematical accuracy and proper procedure. They may also check payroll, timekeeping, and billing records. A specialty for auditing clerks today is to verify hospital bills.

An appraiser talks to two sisters during a taping of the *Antiques Road Show* about the dollhouse their grandfather completed in 1939. He determined the dollhouse's value was between $5,000 and $8,000.

estimator-in-training would bring home about $20,000 a year until trained.

Outlook

The outlook for this field is very good for those number-lovers who are also computer-skilled. Before any structure is built or remodeled, the project needs a thorough cost analysis.

FOR MORE INFORMATION

ASSOCIATIONS

American Society of Appraisers
555 Herdon Parkway, Suite 125
Herdon, VA 20170
(703) 478-2228
Web site:http://www.appraisers.org
This is the association for people who set values for antiques, jewelry, real estate, and other tangible assets. The site offers resources for learning about the business of appraising.

Association for the Advancement of Cost Engineering (AACE) International
209 Prairie Avenue, Suite 100
Morgantown, WV 26501
(800) 858-COST (2678)
Web site: http://www.aacei.org
The professional organization for cost estimators. Check out the Web site for membership information, educational information, links, articles, industry news, and much more.

Professional Construction Estimators Association of America (PCEA)
P.O. Box 680336
Charlotte, NC 28216
(704) 987-9978
e-mail: pcea@pcea.org
Web site: http://www.pcea.org
This professional organization promotes the continuing education of construction professionals and craftspeople, working to create standards among the industry.

Society of Cost Estimating and Analysis (SCEA)
101 S. Whiting Street, Suite 201
Alexandria, VA 22304
(703) 751-8069
e-mail: scea@erols.com
Web site: http://www.erols.com/scea
According to its Web site, SCEA is a "non-profit organization dedi-
cated to improving cost estimating and analysis in government and
industry and enhancing the professional competence and achieve-
ments of its members."

Society for Industrial and Applied Mathematics
3600 University City Science Center
Philadelphia, PA 19104-2688
(215) 382-9800
Web site:http://www.siam.org
This organization is made up of professionals who "use mathematics
to solve real-world problems," through engineering and the computer
sciences. The site has information about careers in math-related
fields, although most of them require advanced degrees.

WEB SITE

EurekAlert!
http://www.eurekalert.org
This is an extensive Web directory maintained by the American
Association for the Advancement of Science. It is a good place to start.

BOOKS

Devlin, Keith. *Life by the Numbers.* New York: John Wiley and Sons, 1998.
This is the companion book to the PBS series of the same name; a
fun book.

Huff, Darrell. *How to Figure It.* New York: W.W. Norton, 1999.
A self-help math book that teaches the basics of cost estimating for
vacations, car maintenance, and other necessities of life.

Paulos, John. *Beyond Numeracy.* New York: Vintage Books, 1992. This chatty book discusses whatever fascinates the author, a mathematician, about numbers, formulas, and the use of mathematics. His enthusiasm is contagious. Paulos has also written *Mathematics and Humor*.

9

INSPECTOR AND TESTER

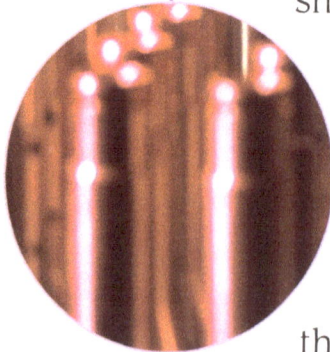

Everything sold in malls, supermarkets, discount centers, and specialty shops is checked over before it leaves the manufacturing plant. All products must meet government standards for safety and durability. Food shouldn't make you sick, gadgets should work in the way they're advertised, and clothes should be well put together. It's up to an inspector to make the decision: "go" or "no go." When a product is defective, the results are frequently highly publicized. No manufacturer wants bad publicity or lawsuits.

The scientific type who is able to make quick decisions is a good inspector or tester. Inspecting takes a good eye for detail, the temperament to perform repetitive tasks, and the desire to do things right. The procedure for testing varies from the simple to the complex. Sometimes an inspector uses only senses—sight, smell, hearing, feel, and taste—but this kind of simplified inspection is less common as sophisticated testing equipment has become miniaturized. Most inspectors today use hand-held scanners and imaging systems.

Inspectors use clearly defined test methods to determine if products meet company and government standards. These standards are usually in writing and constitute a set of requirements to be satisfied by a material, product, or system. Products either pass or fail.

When inspectors judge a product ready to be sold, they often acknowledge it by enclosing a piece of paper with their inspection number in the product box. That way, if the consumer finds something wrong, the inspector can be traced.

Testers do more than inspect. They evaluate the reliability and durability of a product; is it really as good as the

A metal inspector carefully examines parts for flaws.

ads promise? Manufacturers say yes; some consumers' groups may say no. Nondestructive testers (NDTs) test the product without affecting its usefulness. This is a fast-growing field. By using imaging equipment such as ultrasound, an NDT can detect internal or external imperfections in a product, determine its structure and composition, and measure its geometric characteristics. Other emerging equipment for testing products uses neutron beams, magnetic particle tests, and interferometry.

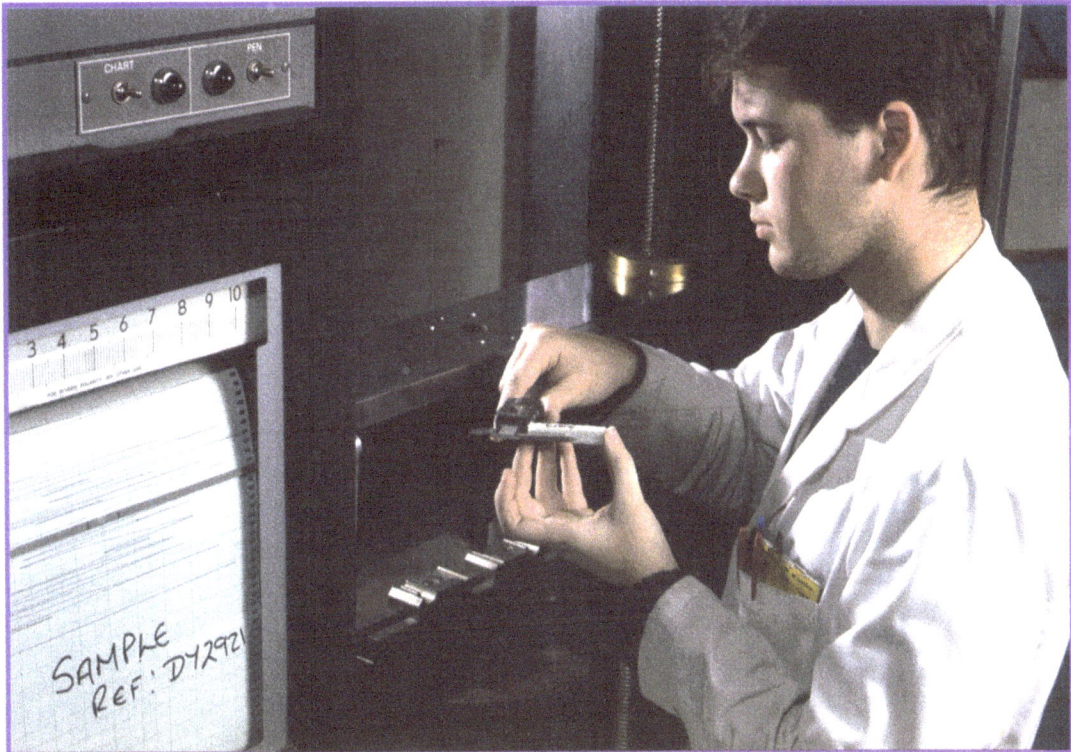

A technician carries out a quality control test using high-tech equipment.

Interferometers measure wave frequency, length, and velocity, and are important in testing electronic equipment.

Another important aspect of testing is checking for safety—trying to ignite flameproof pajamas, for instance, or trying to sink a floatation device. Still another offshoot is calibrating tools to meet universal standards. Without proper testing and calibration, there is no way to be certain that tools are accurate. Being off a fraction of an inch can be critical if the tool is used to work on automobiles or aircraft.

Standards

Most Americans take the use of standards for granted and give no thought to the people who test products. A calendar by the ASTM (American Society for Testing and Materials) may help people appreciate why standards are so important

- **January** Skis, snowboard bindings, ice hockey skate blades, helmets, and face masks all must pass safety inspections before they appear on the slopes or in a rink.
- **February** (Heart month) Defibrillators must always be in a "go" mode and must be perfectly calibrated.
- **March** The "R-value" standard ensures that insulation will resist heat loss, even when March winds blow.
- **April** Earth Day reminds us that standards for clean water and air have made life better for people, animals, and plants.
- **May** Playground equipment must meet standards for safety before a school or park will buy it.
- **June** Bikes, baseballs, mitts, and protective headgear must be approved before sold.
- **July** Airplanes, airports, boats, and cars require a whole army of specialized inspectors.

- **August** Standards for exterior windows, doors, and storm shutters were developed as a result of Hurricane Andrew.
- **September** Labor Day reminds us of the long list of workplace protections that have come about because of labor union trends.
- **October** Parents insist that Halloween costumes and masks meet standards for visibility and being fireproof.
- **November** All of the food on the Thanksgiving table, from turkeys to cranberry sauce, has been inspected and declared healthy before it arrives at grocery stores.
- **December** Safety inspection of toys always gets media attention this month, as consumer groups go beyond the standards set by either the industry or the government.

Education and Training

Training requirements vary, depending on the responsibilities of the employee. Inspectors who give a "pass/fail" grade need little training beyond high school, but this type is being phased out. As testing and inspecting become more automated, technicians will need more training in the use of sophisticated equipment. Much of this training can be done on the job. An associates degree in engineering is recommended for the ambitious tester.

Salary

The median income for inspectors and testers is just short of $12 per hour. Supervisors make more money, but there is little chance for high salaries in this field at present without an engineering degree, either an A.A. or a B.S.

Outlook

There is high turnover in this field, and entry-level positions are usually available. Almost all manufacturers use inspectors and testers. They are most in demand in the aerospace, automotive, electronics, and ordnance (weaponry) industries.

FOR MORE INFORMATION

ASSOCIATIONS

American Society for Nondestructive Testing (ASNT)
1711 Arlingate Lane
P.O. Box 28518
Columbus, OH 43228-0518
(614) 274-6003
Web site: http://www.asnt.org

ASNT is a technical association that encourages the exchange of technical information about testing, offers educational materials and programs on the NDT, and gives certification to qualifying candidates in the field.

American Society for Quality (ASQ)

611 E. Wisconsin Avenue
P.O. Box 3005
Milwaukee, WI 53201-3005
(800) 248-1946
e-mail: cs@asq.org
Web site: http://www.asq.org
From the organization's Web site: "The American Society for Quality (ASQ) has been the leading quality improvement organization in the United States for more than 50 years. Its members have initiated most of the quality methods used throughout the world, including statistical process control, quality cost measurement and control, total quality management, failure analysis, and zero defects."

American Society for Testing and Materials (ASTM)

1016 Race Street
Philadelphia, PA 19103-1187
(215) 299-5400
Web site: http://www.astm.org
This organization sets the technical standards for industries world-wide. Their site contains information about standards and the online magazine, *Standardization News.*

WEB SITES

Nondestructive Testing Information Analysis Center (NTIAC)

http://www.ntiac.com
NTIAC, the Nondestructive Testing Information Analysis Center, maintains this site as a clearinghouse in the field. There are many links and much technical information.

BOOKS

Quality Press Bookstore
The American Society for Quality (ASQ)
611 E. Wisconsin Avenue
P.O. Box 3005
Milwaukee, WI 53201-3005
(800) 248-1946
e-mail: cs@asq.org
Web site: http://www.asq.org
The world's largest publisher of quality-related products offers basic to advanced resources for individuals at various levels of the quality profession.

PERIODICALS

The American Society for Quality (ASQ)
611 E. Wisconsin Avenue
P.O. Box 3005
Milwaukee, WI 53201-3005
(800) 248-1946
e-mail: cs@asq.org
Web site: http://www.asq.org
The ASQ publishes several trade journals, including *Quality Progress, Journal of Quality Technology (JQT), Quality Engineering, Quality Management Journal (QMJ), Software Quality Professional (SQP), Technometrics*, and *The Informed Outlook*.

10

SURVEYING TECHNICIAN

If you like to read maps and can visualize them in three dimensions, you might have the aptitude to be a surveyor. Surveyors measure distances, directions, and angles on the earth's surface. They calculate elevations (mountains) and depressions (valleys). They mark boundaries for landowners, and they record the results of

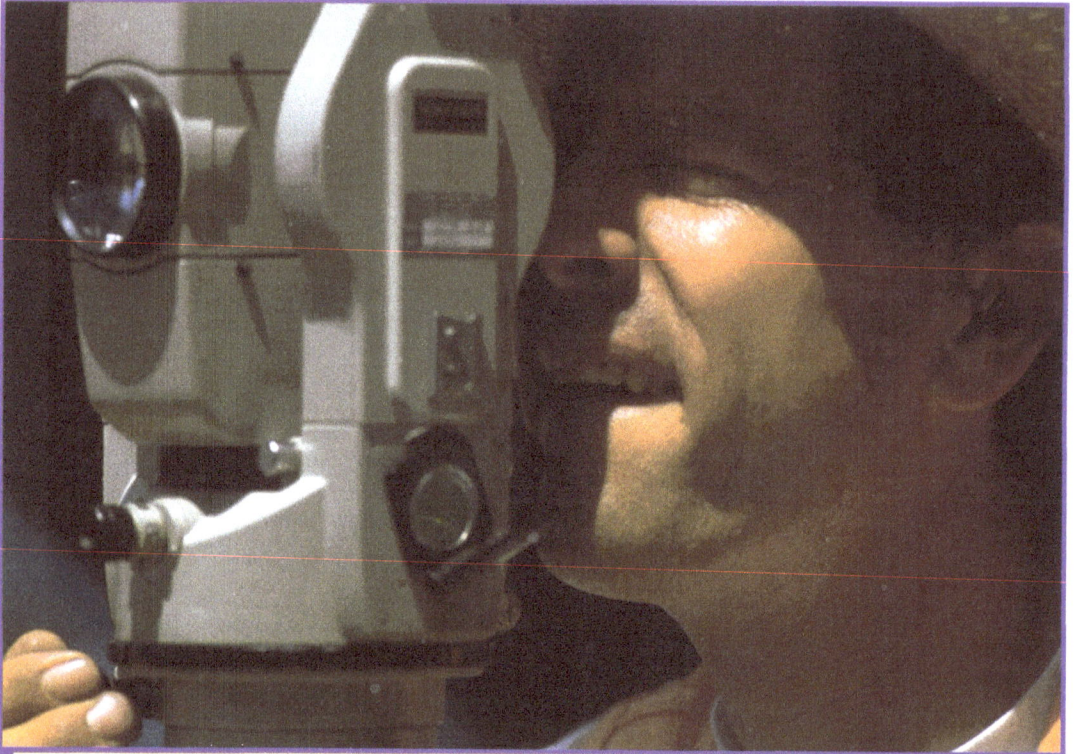

A surveying technician uses sophisticated equipment to measure a tract of land.

their surveys for cartographers. Before a new housing development, shopping center, or sports arena can be built, the land must be surveyed.

The most important tool for surveyors is geomatics—a combination of geometry and mathematics. Geomatics software has changed surveying dramatically in the last ten years. Today's surveyors spend more time in front of their computers than in the field. Simulation programs create virtual landscapes, and information from satellites (teledetection) fills in the gaps. Teledetection works via GPS, the

Global Positioning System. GPS is made up of a group of twenty-four satellites orbiting Earth. From their signals, it is possible to pinpoint the exact location of anyone or anything containing a remote sensing device. These devices appear routinely in new cars and are small enough to be carried by hikers.

EROS, Earth Resources Observation System, is the ambitious program from the U.S. Geological Survey (USGS) and NASA. By way of GPS satellites, USGS is surveying and mapping the entire earth, with the cooperative efforts of surveyors and cartographers worldwide. A key element in this project is the satellite *Landsat7*. The data collected can be searched online through Earth Explore (a fee-based service). It can be accessed quickly in times of emergency, such as a flood, volcano eruption, or major oil spill.

NASA oversees a similar, more comprehensive, program called EOS, Earth Observing System. This project tracks major concerns of earth scientists, such as global warming and holes in the ozone layer. It studies the earth as a complete environmental system.

Photogrammetry is an important part of both surveys. It is the piecing together of aerial photographs—some from airplanes, some from satellites—to reconstruct the geography of an area. Photogrammetry is a tool used frequently by survey technicians.

Two Surveyors

Steve Perron, a Canadian cartographic technician, uses his skills in surveying to map the forest areas in Quebec for the Ministry of Natural Resources. He works as part of a team with the goal of monitoring forest growth and assessing resources. He relies heavily on his computer, using geomatics software and extracting information from a database.

Another Canadian, David Cody, is mapping the coast of Nova Scotia for the government. Using photogrammetry, he is producing 117 maps of the coastal waters belonging to Canada to the twelve-mile limit. Water mapping, or hydrography, is important to the maritime provinces, which rely heavily on fishing.

More information on these jobs can be found in the Canadian online booklet, "Look Ahead, Get Ahead" via the Human Resources Development Web site, http://www.cthrb.ca/.

When they are in the field, survey technicians work as part of a team under the direction of a licensed surveyor. They use traditional equipment, such as a theodolite (an instrument with a telescopic site that measures horizontal and vertical angles), and make notations. Later, they use a

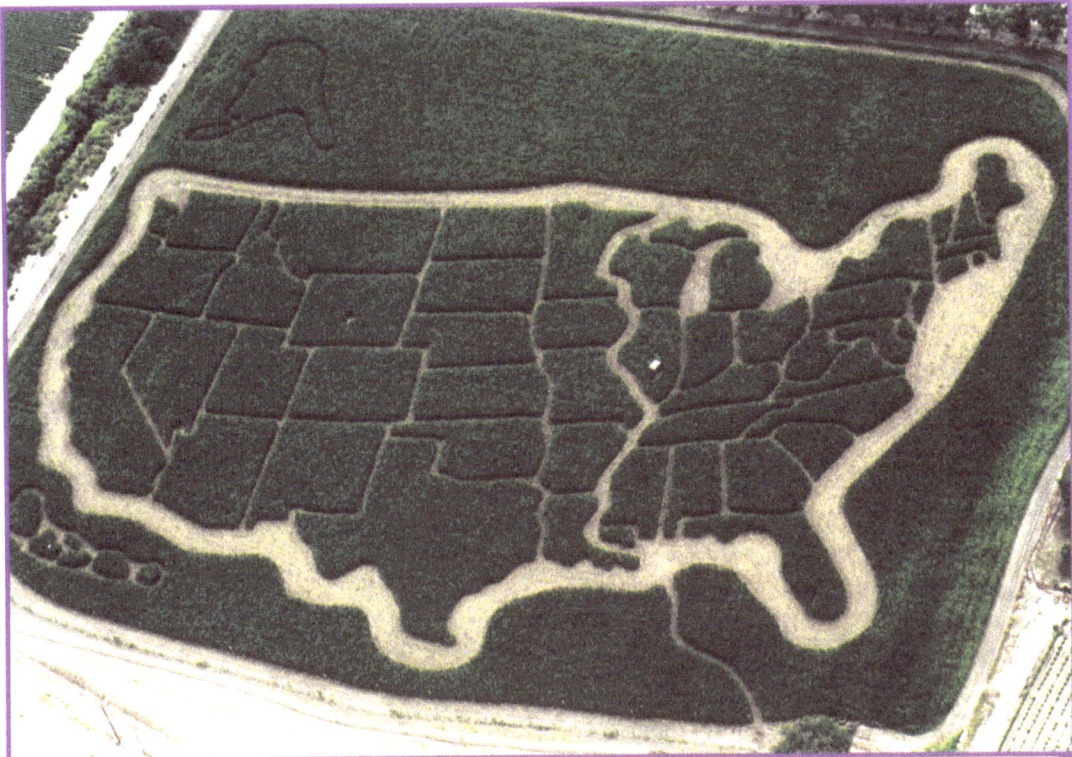

The Mesilla Valley Maiz Maze, a twelve-acre cornfield cut into a maze of the United States of America, was made with the help of a global positioning system receiver.

CAD system to plot coordinates and draw a map of what they've measured.

Surveying jobs are most frequently found with national, state, provincial, and local government agencies, such as the U.S. Geological Survey (USGS), Bureau of Land Management (BLM), Canadian Ministry of Natural Resources, U.S. Army Corps of Engineers (USACE), U.S. Forest Service, National Oceanic and Atmospheric Administration (NOAA), and the National Imagery and Mapping Agency (NIMA). Surveying technicians can also find work in construction firms, with mining, oil and gas companies, and with public utility companies. Those who choose government service must take a civil service exam or the equivalent before being hired.

Education and Training

High school students interested in surveying should take courses in algebra, geometry, trigonometry, drafting, mechanical drawing, and computer science. With such a background, they may start as apprentices. With postsecondary training, they can start as technicians or assistants. With on-the-job training and some formal training—either in institutions or by correspondence—workers may advance to other positions. CAD instruction is essential. It is available in most high schools, vocational-technical schools, and community colleges.

Salary

The starting pay for surveyor technicians averages $12 per hour in the private sector and $13.50 per hour in government service. It is possible to advance with experience and the assumption of responsibility.

Outlook

There will continue to be a moderate demand for surveyors skilled in using geomatics software and geographic imaging systems.

FOR MORE INFORMATION

ASSOCIATIONS

American Congress on Surveying and Mapping (ACSM)
6 Montgomery Village Avenue, Suite #403
Gaithersburg, MD 20879
(240) 632-9716
Web site: http://www.acsm.net
The ACSM is a non-profit organization encompassing four member groups: the American Association for Geodetic Surveying (AAGS), the Cartography and Geographic Information Society (CAGIS), the Geographic and Land Information Society (GLIS), and the National Society of Professional Surveyors, Inc. (NSPS).

ASPRS: The Imaging and Geospatial Information Society
5410 Grosvenor Lane, Suite 210
Bethesda, MD 20814
(301) 493-0290
e-mail: asprs@asprs.org
Web site: http://www.asprs.org
According to its Web site, the ASPRS seeks to advance knowledge and improve understanding of mapping sciences and to promote the responsible applications of photogrammetry, remote sensing, geographic information systems (GIS), and supporting technologies.

North American Cartographic Information Society (NACIS)
P.O. Box 399
Milwaukee, WI 53201-0399
(414) 229-6282
Web site: http://www.nacis.org
The professional organization for mapping specialists.

The Remote Sensing and Photogrammetry Society

Web site: http://www.rspsoc.org

This association provides a forum "for the meeting of academics, commerce, and government." There are many links to remote sensing sites hosted by government agencies and universities and a listing for jobs worldwide.

WEB SITES

CAD/CAM World

http://www.geocities.com/cadcamworld/

This is a megadirectory with more than 300 sites relating to CAD/CAM. It evaluates software, hardware, and books relating to the field.

Earth Observing System (EOS)

http://eospso.gsfc.nasa.gov

This site has information about the NASA project for the twenty-first century, Mission to Planet Earth. The Earth Observing System via GPS is an important part of this mission.

Geological Survey of Canada

http://www.nrcan.gc.ca

This site features a bookstore with books relating to surveying and links to other geoscience organizations.

U.S. Geological Survey

http://www.usgs.gov

This is a global site with information on earthquakes, flooding, and other upheavals on the planet, plus mapping and remote sensing. Links will take you everywhere on, under, and above the surface of the earth. From this site you can find the EROS (Earth Resources Observation System) site.

BOOKS

Jones, Christopher. *Geographical Imaging Systems and Computer Cartography*. Reading, MA: Addison-Wesley, 1997.
A look at the techniques involved in the graphic display of spatial information, also known as mapping.

Kraak, Menno-Jan, and Ferian Ormeling. *Cartography: Visualization of Spatial Data*. Reading, MA: Addison-Wesley, 1996.
This is one of the standard books in the field. It is an introduction to map scale, projecting, and map graphics.

CD-ROM

A tutorial on remote sensing is available from NASA for $10.
Nancy Dixon
NASA/Goddard Space Flight Center
Code 835, Bldg. 28, Room W186
Greenbelt, MD 70770

11

FIBER-OPTIC
TECHNICIAN

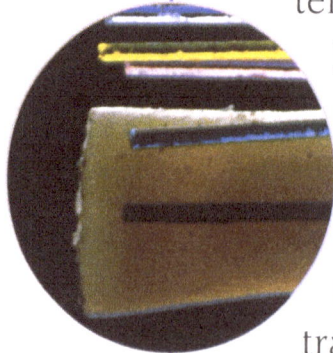

Fiber optics is said to be revolutioniz-
ing communications much as the
telephone did 100 years ago. If
you want to be on the cutting
edge of technology, this is a
good place to be. The
"optics" come from a con-
centrated beam of light, a
laser, which can be used to
transmit voice, data, and video
at the same time—a billion or more

bits of information per second. The "fiber" is a bundle of transparent, hair-thin flexible strands of glass. Fiber cables are smaller and lighter than conventional cables using copper wires or coaxial tubes, yet they carry much more information and the flow of information proceeds more smoothly.

The combination of laser and glass offers a fast, reliable, and inexpensive means of telecommunication. Because of its advantages over metal wiring, optical fiber is rapidly replacing wire in computers, photocopiers, medical equipment, navigation guidance systems, and weaponry.

Optical fibers are immune to interference from lightning or nearby electric motors, so there is no interruption in the television signal when someone runs a disposal in the kitchen or during a storm. Optical fibers don't "get their wires crossed." There's no bleed-through of the voices talking on other lines during a telephone conversation. Eavesdropping on conversations or bugging is harder to do and easier to detect.

Fiber-optic technicians assist engineers in designing and testing new uses for fiber optics. They set up electrical and electronic experiments, which may lead to the development of new applications. The technology is still new enough to make experimenting with it exciting. We don't yet know the extent of what fiber optics can do or where it will lead.

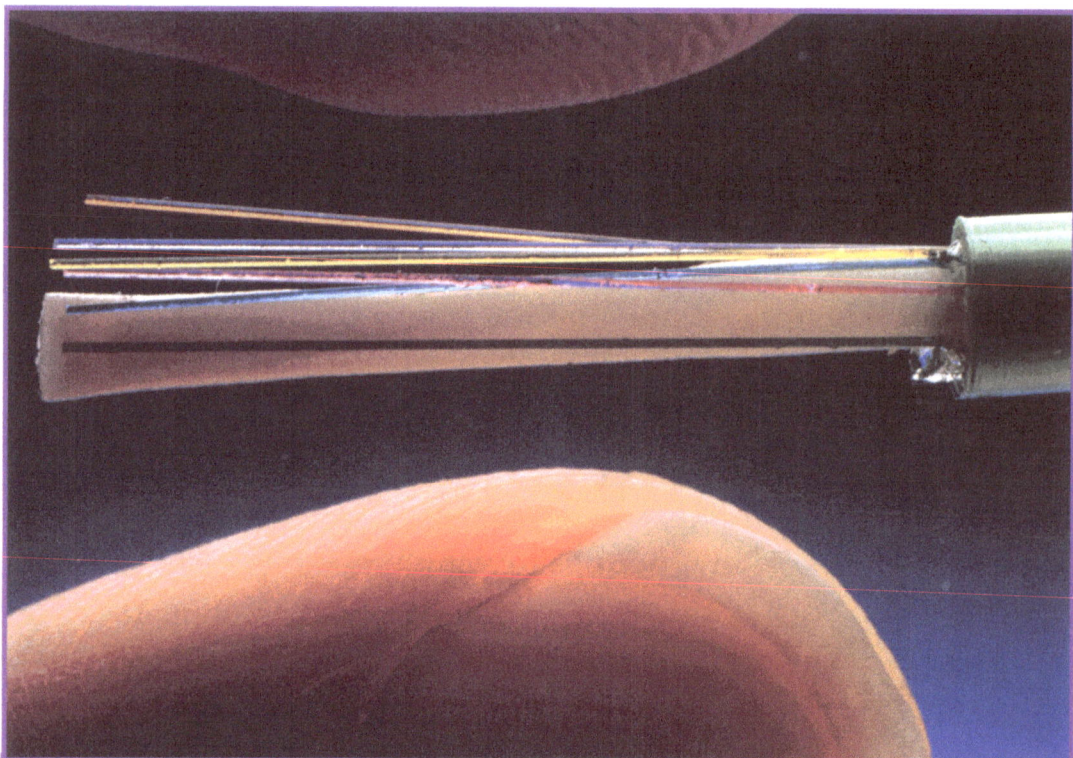

With the United States currently being "rewired" with fiber-optic cables, opportunities for fiber-optic technicians will continue to grow well into the future.

Some technicians repair and install existing laser and optical fiber devices and systems. These systems are most often in telephone and computer networks today. Using spectrometers, they measure light frequencies emitted by lasers and adjust them when necessary.

The most active field is designing fiber-optic systems that link buildings and people. For instance, some school systems are converting to fiber-optic networks so that they have rapid access to multimedia presentations coming from the information superhighway. Figuring out where cable hookups should go and splices should be made is up to technicians. Splicing is done with high-powered lasers that fuse cables and require that technicians be on-site.

Experts predict that creating such networks will keep fiber-optic technicians busy for many years to come. New uses, such as in surgery or engineering, will open new fields for employment within the next decade. Laser surgery has long been in practice; now fibers are also being used. The automobile and aviation industries are already investigating how they can best use fiber optics.

Education and Training

Many companies provide on-the-job training to high school graduates who have a strong background in math and physics. Community colleges and technical institutes offer programs in the field. The armed forces also offer training in laser installation and operation. There are also online training courses in fiber optics which are endorsed by fiber-optic associations and can lead to certification. As in all technological fields, the technician must be prepared to keep learning.

Salary

The starting salary for a fiber-optic technician is about $18,000 per year. It can rapidly rise to near $40,000, but

A Trillion Bits a Second?

A trillion bits (units of information) a second is a tremendous pipeline. It's more than all the long-distance traffic that usually flows in the whole American telephone network. It's the equivalent of 10 million voice telephone circuits. A trillion-bit fiber could carry a mind-numbing hundreds of thousands of standard television channels. This is what fiber-optic developers are aiming for.

What else is in the future? Networks that route signals by wavelength, or color. Chicago might be violet, Cincinnati orange, Pittsburgh green. Continents will be color coded, as will nations within continents, and cities within nations. Rainbows of light will spread around the globe faster than anyone can blink.

—Adapted from *City of Light: The Story of Fiber Optics* by Jeff Hecht (Oxford University Press, 1999).

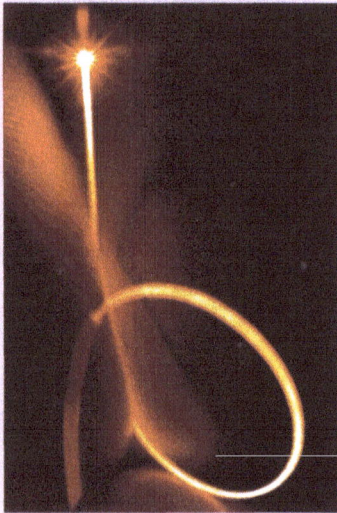

A single strand of optical fiber can handle five million telephone conversations at once.

Western Energy Services linemen secure a section of fiber-optic line to a utility pole in Farmington, New Mexico.

because the field is in a growth mode, it is hard to predict what salaries will be in five or ten years.

Outlook

Right now there is a shortage of fiber-optic technicians and the job outlook is excellent for the next five years or so. Jobs are primarily in electronics and telecommunications today, but many manufacturing firms are converting to laser-driven systems, so the job market will expand in the years ahead. The opportunities for fiber-optic technology in medicine, aviation, and other forms of transportation have only been hinted at.

FOR MORE INFORMATION

ASSOCIATIONS

The Fiber Optic Association (FOA)
Box 230851
Boston, MA 02123-0851
(617) 469-2362
e-mail: info@thefoa.org
Web site: http://world.std.com/~foa/
FOA is an international professional society for the fiber-optic industry. Its aim is to develop educational programs, certify fiber-optic technicians, and provide lists of approved fiber-optic training courses. The site has information about testing for certification and job opportunities. There is an online newsletter.

Laser Institute of America
13501 Ingenuity Drive, Suite 128
Orlando, FL 32826
(407) 380-1553
Web site: http://www.laserinstitute.org/
The Laser Institute is a professional society which emphasizes safety in laser research. It has an online bookstore and job board.

Optical Society of America
2010 Massachusetts Avenue NW
Washington, DC 20036
(202) 223-8130
Web site: http://www.osa.org
Optics is the science of light. This Web site has an infobase, book-store, and links to other science and technology sites.

WEB SITES

EurekAlert
http://www.eurekalert.org
This is a mega Web directory maintained by the American Association for the Advancement of Science. It is a good place to start.

Fiber Optics Online
http://fiberopticsonline.com
This commercial site offers a large career development section with a job search and recruitment center, in addition to news and products.

Human Resources Development Canada
http://hrdc-drhc.gc.ca
Canada's Human Resources Development Agency has an extensive and easy-to-use Web site on technology jobs, including learning and training resources, career planning tools, and self-assessment and aptitude testing.

Optics.org
http://optics.org
This site is the highest of high tech. For those who can speak the language, there is a searchable database of jobs and information in this cutting-edge field.

Women in Technology International (WITI)
http://www.witi.org
This organization offers special encouragement to women who want to succeed in technology.

BOOKS

Crisp, John. *Introduction to Fiber Optics*. Boston: Newnes, 1996.
A good foundation for understanding how fiber optics works.

Goff, David R. *Fiber Optic Reference Guide: A Practical Guide to the Technology.* Boston: Focal Press, 1999.
A broad reference book that teaches the fundamentals of fiber optics.

Hecht, Jeff. *City of Light: The Story of Fiber Optics*. New York: Cambridge University Press, 1999.
This huge book is a comprehensive, highly technical account of the development of fiber optics.

Palais, Joseph C. *Fiber Optic Communications.* Upper Saddle River, NJ: Prentice-Hall, 1998.
A great introduction to fiber optics.

12

ELECTRONICS TECHNICIAN

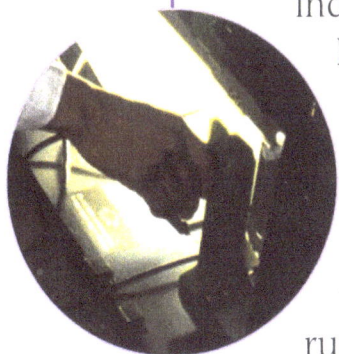

Electronics make the world go 'round. We are now so dependent on this industry that we can't imagine life without it. Electronics uses power that comes from electrons, tiny (subatomic) particles that carry a negative electrical charge. Circuits are routes for these electrons to run, beginning and ending at the same point. When many circuits are put together on the same chip or wafer, they are called integrated circuits. This technology has eliminated the need for wiring and allowed for miniaturization, which means that the radio console that took up a corner of the living room fifty years ago can now fit in your pocket.

Switching and timing circuits, operated by microprocessors, opened the world to telecommunications. Electronic telecommunications now can connect us to everyone on the planet. Its equipment includes much more than just telephones. Radios, fax machines, scanners, garage door openers, alarms, computers, televisions, VCRs, CD players, DVDs, video cameras, surround-sound systems, and some high-tech weapons are classed as telecommunications because information is sent and received through them.

Electronics technicians install or repair such equipment. There are two methods of repairing. Bench technicians work in a lab or shop. The malfunctioning equipment is brought to them for repair. Field technicians, on the other hand, go to the site of the problem (office or factory) and bring their diagnostic equipment with them. Large machines, such as photocopiers, cannot easily be transported and must be fixed on site. Smaller ones, such as fax machines, are usually sent back to the manufacturer for repairs.

Diagnostic equipment includes multimeters, which measure the voltage and resistance of the power supply;

An electronics technician repairs a copy machine. Electronics technicians must often travel to their clients' locations to provide their services.

color bar and dot generations, which provide on-screen test patterns; signal generators, which provide test signals; and oscilloscopes, which measure waveforms.

The trend for the last several years has been to replace failing equipment, rather than fix it, except in the case of expensive items still under warranty or a service contract. Because each year's models have been an improvement over the previous year's, it makes more sense to buy new than to repair old. Also, much electronic equipment is

An electronics technician tests a part to find out why it's not working properly.

self-diagnosing, even self-repairing, making less need for repair people.

Telecommunications technicians also work within corporations to set up communications systems using cell phones, voice mail, auto response, call accounting, and interfacing with computer systems. Computer network technicians are specialists in installing and repairing local area networks (LANs) that connect all phases of a business with the corporate headquarters. These technicians may become certified netware engineers (CNEs) after a short course of study.

Electronics technicians may work in research and development (R&D) labs with scientists and engineers who design and test new equipment. There are hundreds of electronics firms in North America that maintain labs and hire technicians. With a background in repair, technicians can become excellent troubleshooters and, using sophisticated diagnostic equipment, find problems quickly. When they spot potential problems in new products, they can debug them and make the products trouble-free.

An avenue that some electronics technicians are now exploring is hybrid electronics. Instead of using standard boards imprinted with circuits, hybrid electronics uses printed ceramic microcircuits with a layer of film. There seem to be some advantages to this process, but so far the field is a small one.

Electronics technicians should understand the basic laws of electricity and electrical circuits and the principles of electronics. They also must be able to read schematics. They must be able to distinguish colors (wires are color-coded) and tones (as in a touch-tone phone), so it is not a field for the color-blind or tone-deaf.

Education and Training

Training in electronics or telecommunications is available through community colleges, technical institutes, the armed

An electronics technician with the North Carolina Department of Transportation cleans traffic light reflectors as part of a preventive maintenance routine.

forces, correspondence courses, and via the Internet. Equipment manufacturers and software developers also provide training, and formal apprenticeship programs have been established. It is wise to aim for certification in this field. Certification is available from ISCET (the International Society of Certified Electronics Technicians) and ETA (Electronics Technicians Association).

Salary

The wages for electronics technicians vary widely. The starting rate for a repair shop job is about $10 per hour. An experienced electronics technician who works for a large company, whether in repair or installation, can expect to earn an annual salary of close to $50,000. For those technicians who join a union, the benefits can be considerable.

Outlook

This field is expected to grow at an average rate through 2010, not as quickly as it did in the 1990s. It is almost guaranteed to change in the next ten years, which will probably spur growth in certain areas. The need for installers will increase, and the need for repairers will decline. Jobs for electronics technicians are mainly in private industries, especially the giants in the communications industry.

FOR MORE INFORMATION

ASSOCIATIONS

Communications Workers of America
501 3rd Street NW
Washington, DC 20001
Web site: http://cwa-union.org
(202) 434-1100
CWA, a trade union, offers training programs and a few scholarships. There is information about bilingual opportunities.

Electronics Technicians Association
502 N. Jackson
Greencastle, IN 46135
Web site: http://www.eta-sda.com
This site has links to training programs that lead to certification.

International Brotherhood of Electrical Workers (IBEW)
Telecommunications Department
1125 15th Street NW, Room 807
Washington, DC 20005
Web site: http://ibew.org
This site contains information on union activities and benefits, and has an online magazine.

International Society of Certified Electronics Technicians (ISCET)
3608 Pershing Avenue
Fort Worth, TX 76107-4527
(817) 921-9101
Web site: http://www.iscet.org
ISCET administers certification programs. The applicant must pass qualifying exams to be certified.

Telecommunications Industry Association
2500 Wilson Boulevard, Suite 300
Arlington, VA 22201
(703) 907-7700
Web site: http://www.tiaonline.org
This site is from the point of view of the industry, with information on trade shows and what's new in the marketplace.

United States Telecom Association
1401 H Street NW, Suite 600
Washington, DC 20005-2164
Web site: http://www.usta.org
This organization is an association of independent telephone companies. There is information on training and employment at this site.

WEB SITES

America's Learning Exchange
http://alx.org
America's Learning Exchange's site contains information on training offered on the Net and various certification programs.

Bureau of Apprenticeship and Training
http://bat.doleta.gov
Provides information on apprenticeships in the United States.

Women in Technology International (WITI)
http://www.witi.org
This organization offers special encouragement to women who want to succeed in technology.

BOOKS

Amdahl, Kenn. *There Are No Electrons.* Arvada, CO: Clearwater Publishing Co., 1991.
An off-the-wall, even funny, explanation of electronics in theory and practice.

Maxfield, Clive Max. *Bebop to the Boolean Boogie: An Unconventional Guide to Electronic Fundamentals, Components, and Processes.* Solana Beach, CA: HighText, 1995.
Reviewers call this book "clear, comprehensive, funny, and informative." It's a technical introduction to electronic circuitry than doesn't overwhelm the reader with detail.

PERIODICALS

Electronics Weekly
125 Park Avenue
New York, NY 10017
(212) 370-7440
Web site: http://www.electronicsweekly.co.uk
Electronics Weekly provides all of the latest news and features regarding the electronics world.

Electronics World
125 Park Avenue
New York, NY 10017
(212) 370-7440
Electronics World is a print publication serving electronics consumers and professionals.

13

CHEMICAL TECHNICIAN

Chemistry is no longer the glamour field it once was, but good jobs are still available. If you enjoy concocting, brewing, cooking, and experimenting, chemistry is a field to investigate. There are two basic types of chemistry: organic and inorganic. Organic chemistry is based on compounds containing carbon. They make up most of commercial chemistry. Inorganic chemistry is concerned with the elements, metallic and nonmetallic.

Most chemical technicians work in research and development (R&D) labs for large multinational corporations that process food, make plastics, develop medicines, produce commercial cleaning products, or manufacture industrial and agricultural chemicals. These companies fund research in the hope of striking it rich with a new, profitable product.

Chemical technicians assist professional chemists in analyzing complex substances and running experiments on new combinations of chemicals. As in any recipe, some combinations are better than others. A cosmetic that makes a woman look beautiful but smells like decaying garbage is not going to sell. Chemists think that their failures are as important as their successes. Failures tell them what not to do.

An important part of the organic chemistry business is the quest for new ways to synthesize products. This is especially true in the pharmacy industry. A naturally occurring antibiotic may be very expensive to produce and may not be able to be produced in large quantities. If scientists can determine the structure of molecules in the antibiotic and copy it synthetically, the company can mass-produce it, bringing down the cost to consumers. While they are synthesizing molecules,

A chemical technician must keep careful records on the hundreds of substances he tests.

chemists can also modify them, eliminating undesirable side effects and enhancing their power.

Chemical technicians are often responsible for setting up experiments and taking care of equipment, which may be as simple as glass beakers and pipettes, or as complex as gas analyzers that perform several functions at once. They must keep everything clean and in good working order.

Technicians must be math and computer literate. Using formulas, they measure the characteristics of liquids and gases—volume, density, temperature, and volatility (the speed of evaporation). After carefully recording the results, they present their findings and conclusions to chemists.

Organic chemical technicians are also in demand in environmental agencies, where they monitor and assess water and air quality. This is a growing field and will become more specialized in the years ahead.

Inorganic chemical technicians are likely to work in heavy industry, in metals-testing laboratories or metals-processing plants. Some industrial chemicals are inorganic, such as battery acid or various salts. Industrial ceramics is beginning to attract the attention of scientists and industry and is predicted to be a growing field over the next ten years, as will the highly specialized area of superconductive metals—metals used in electronics applications.

Education and Training

Two-year programs leading to an A.A. are available in many community colleges. This is recommended to anyone seeking a quality job in chemistry, although it is not necessary for entry-level positions. It is crucial to take as much math as possible. Most employers offer on-the-job training and some require that applicants take a test to show their level of proficiency. While a background in general chemistry is important, it is wise to specialize, first in organic or inorganic

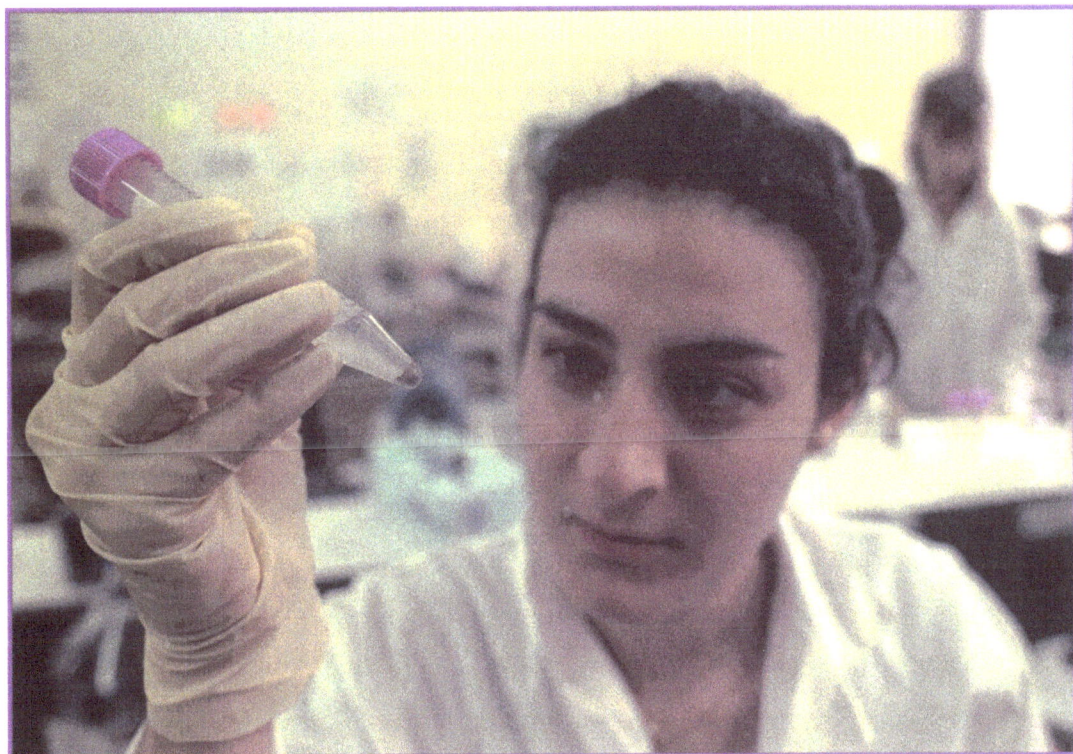

A laboratory technician tests water samples for signs of pollution.

chemistry, and then narrowing your focus even further. Specialists can command the best jobs, both in terms of salary and in responsibility and working conditions.

Salary

A chemical technician can expect to start at around $20,000 per year and work up to $30,000. The technician who can specialize and who continues learning through formal classes or conferences can expect to make more.

Help Wanted: Chemistry Technician

Employees prepare and validate samples and solutions used for testing organic and/or inorganic specimens. They calibrate and adjust chemical instruments such as spectrometers, auto analyzers, gas chromatographs, and atomic absorption spectrometers to ensure accurate results. Employees record the data obtained and perform necessary mathematical calculations. Work is supervised by a professional scientist.

An associate's degree in chemistry or nine hours of college chemistry plus related experience is required. Applicants must pass a three-hour written test covering chemical processes, arithmetic calculations, proficiency in reading and interpreting technical materials, and identification of laboratory glassware.

Outlook

Right now, the major chemical companies are merging and retrenching, and economists do not expect much growth in the years ahead. Two exceptions are the pharmaceutical companies and environmental services. There is no doubt that these areas will continue to grow. The chemical industry is a good example of globalization; it reaches everywhere on the earth, regardless of national boundaries. Employees who can speak more than one language are at an advantage.

FOR MORE INFORMATION

ASSOCIATIONS

American Chemical Society
1155 Sixteenth Street NW
Washington, DC 20046
(202) 872-4600
Web site: http://www.acs.org
This organization is very active in promoting chemistry and supporting chemists. The Web site is extensive. It has detailed descriptions of jobs in various chemical fields, from agriculture to water, classified job ads, information on education, an online reference library, and an online news magazine.

SOCMA (Synthetic Organic Chemical Manufacturers Association)
1850 M Street NW, Suite 700
Washington, DC 20036
(202) 721-4100
Web site: http://www.socma.com
This site, maintained by chemical manufacturers, offers much information on the industry and includes employment opportunities.

WEB SITES

America's Learning Exchange
http://alx.org
America's Learning Exchange's site contains information on training offered on the Net and various certification programs.

Bureau of Apprenticeship and Training
http://bat.doleta.gov
Provides information on apprenticeships in the United States.

Chemical Periodic Table
http://www.chemicool.com
The Periodical Table of elements can be admired at this site.

Links for Chemists
http://www.liv.ac.uk/Chemistry/Links/intro.html
Links for Chemists is a megasite with links to over 8,400 chemistry resources on the Web. It is maintained by the University of Liverpool in the United Kingdom.

Women in Technology International (WITI)
http://www.witi.org
This organization offers special encouragement to women who want to succeed in technology.

BOOK

Karakstis, Kerry, and Gerald R. Van Hecke. *Chemistry Connections: The Chemical Basis of Everyday Phenomena.* New York: Harcourt, 1999. The information in this illustrated book is presented in an easy-to-understand, question-and-answer format.

PERIODICALS

American Scientist
P.O. Box 13975
Research Triangle Park, NC 27709-3975
(919) 549-0097
Web site: http://www.amsci.org
This bimonthly magazine features reviews of current research in the scientific field.

Chemical Engineering
110 William Street
New York, NY 10038
(212) 621-4900
Web site: http://www.che.com
A magazine published for those interested in the chemical process industries.

Chemical Week
110 William Street
New York, NY 10038
(212) 621-4900
Web site: http://www.chemweek.com
A consumer magazine catering to professionals in the field of
chemistry and other sciences.

14

CIVIL ENGINEERING TECHNICIAN

Engineers are practical scientists. They solve matter-of-fact problems. They design, build, and repair things, from bridges to paper-clips, DVDs to anti-lock brakes. Engineering technicians assist engineers and scientists, especially in re-search and development. Engineering teams are employed by most large manufacturers to build prototypes of new products and test them for safety and reliability. Technicians are employed in all fields of engineering and work under the careful supervision of engineers.

Civil engineering is a discipline that is a marriage between technology and art, or architecture. It is responsible for all the great structures of the world: skyscrapers, sports domes, dams, and bridges. Architects design buildings and advise how they should be built. Civil engineers (CEs) wear hard hats and oversee the construction. They have to know which ideas will work and which won't, before millions of dollars are spent.

While civil engineers assess strength of materials, stress loads, traffic flow, and use patterns, they also look at form and design. They like efficiency and practicality. Ideally, they create something useful that is also attractive, like a cloverleaf intersection in a highway.

Technicians work with civil engineers and architects to plan and build highways, buildings, bridges, dams, wastewater treatment systems, airports, and other large facilities. Civil engineering work is most in demand in urban areas, so the employment of CEs and technicians usually follows population growth. Their skills in building infrastructures have allowed for the growth of cities.

Civil engineering technicians are involved in assessing the feasibility of complex structures such as the Syracuse University Carrier Dome.

Technicians frequently go with engineers to the site of the new structure, then return to the office to help draw up a proposal. They work with cost estimators to figure a budget for the project. One of the technician's most important tools is the computer, especially CAD systems. CAD programs today allow a technician to "build" a structure completely, then modify it to suit new ideas and needs as they emerge. As software becomes more sophisticated and complex, technicians must keep up with the upgrades.

Numerical values are input, and formulas are calculated, resulting in determination of the materials needed.

In the case of building a bridge, civil engineers figure the weight it must bear, the amount of traffic it should hold, the wind and freezing temperatures it will likely endure, and the other stresses on it to determine the number of supports it will require and its shape and style. They must be able to foresee problems before the structure is built. To do so, they consider a hundred "what if" situations—can the structure withstand an earthquake, tornado, or prolonged subzero temperatures?—even if they are unlikely to occur.

A growing part of civil engineering work is materials testing with imaging systems, such as ultrasound and magnetic particle tests. With imaging systems, they can assess the stress on structures without having to wait until they crack or buckle.

Civil engineers and technicians must work with a host of people involved in the construction projects—clients, architects, government inspectors, surveyors, cost estimators, materials suppliers, labor representatives, subcontractors—so their people skills should be as carefully developed as their math and science skills. The ability to communicate clearly is a must.

Bridge Building

Remember how much fun it was to build things out of Legos and popsicle sticks when you were a kid? A few lucky contestants can do it again. The Illinois Institute of Technology sponsors a yearly bridge-building contest for high school students across the globe.

Each year, the engineering department at IIT draws up a different set of specifications for a bridge and sends them to participating regions. The materials they choose are inexpensive and easy to obtain. The design is up to participants and their model must pass a rigorous test: remain standing after twenty-five kilograms of pressure is applied. (It is suggested that bridges not have far to fall.) Regional winners may advance to the international contest, which is held in a different host city each year.

For more information, check out the Web site: http://iit.edu/~hsbridge.

The Golden Gate Bridge, California

Education and Training

Math and physics are required for all engineering technicians. Those wanting to become civil engineering technicians should enroll in a two-year course at college or technical school, which emphasizes structural engineering. Some of the typical courses are methods of construction, materials testing, strength of materials, traffic engineering, and reinforced concrete design.

Salary

The industry average for all engineering technicians is $35,000 per year. Civil engineering technicians with large architectural firms may expect to make more.

Outlook

The job opportunities for civil engineering technicians are expected to increase over the next ten years. Competitive pressure will force manufacturing facilities and product designs to change more rapidly than in the past. Population growth will require more highways, airports, and water treatment facilities.

FOR MORE INFORMATION

ASSOCIATIONS

Accreditation Board for Engineering and Technology, Inc.
111 Market Place, Suite 1050
Baltimore, MD 21202
(410) 347-7700
Web site: http://www.abet.org
Find out about how to get accredited.

American Society of Certified Engineering Technicians (ASCET)
P.O. Box 1348
Flowery Branch, GA 30542
(770) 967-9173
Web site: http://www.ascet.org
ASCET has a large employment data bank for its members and offers *CET* magazine.

American Society of Civil Engineers
1801 Alexander Bell Drive
Reston, VA 20191-4400
(703) 295-6000
Web site: http://asce.org
This well-designed Web site shows the enthusiasm civil engineers have for their profession. There is information about careers in the field and an online magazine, *Civil Engineering*.

Junior Engineering Technical Society (JETS)
1420 King Street, Suite 405
Alexandria, VA 22314-2794

(703) 548-5387
Web site: http://www.jets.org
JETS' mission is to guide high school students toward their career goals. They offer a self-administered engineering aptitude test; a competition, Tests of Engineering Aptitude, Mathematics, and Science (TEAMS); and a national engineering design competition. (Enclose $3.50 for a full package of guidance materials and information.)

National Institute for Certification in Engineering Technology (NICET)
1420 King Street
Alexandria, VA 22314-2794
(888) 476-4238
Web site: http://www.nicet.org
NICET's site has extensive information on testing for certification (Job Task Competency Exam), locations of testing centers, and exam dates.

WEB SITES

Brantacan
http://brantacan.co.uk/bridges.htm
This is a fun site, and dramatic, too. It contains a collection of photographs of beautiful bridges with explanations on how they were built.

The Bridge Site
http://www.bridgesite.com
This Web site offers information for anyone interested in the construction of bridges, including professional associations, industry events, photo galleries, and discussion forums.

The Civil Engineer
http://www.geocities.com/zekkos_gr/
A site with hyperlinks to categories of civil engineers, such as structural, geotechnical, hydraulic, environmental, and transportation. The online magazine, *CE News*, is available here.

School of Civil and Environmental Engineering
http://www.ce.gatech.edu
This site, maintained by Georgia Tech, is a directory of everything connected to civil engineering. The virtual library has links to journals and articles of special interest, such as pavement engineering.

BOOKS

Beall, Michael E., et al. *Inside AutoCAD 14.* Indianapolis, IN: New Riders Publishing, 1998.
This book gives a step-by-step introduction to this indispensable technology.

Career Information Center. *Engineering, Science, and Technology.* Vol. 6. New York: Macmillan Library Reference USA, 2000.
A handy reference book for those interested in pursuing careers in science and technology. Valuable information about salary and forecast is provided.

GLOSSARY

abatement Lessening the degree of a toxin's presence.

avionics Aviation electronics.

CAD Computer-aided design.

effluent Wastewater discharge.

geomatics A combination of geometry and mathematics.

hydroponic gardening Growing plants without soil.

photogrammetry The process of making maps or scale drawings from aerial photographs.

schematic A structural diagram.

spectrometer An instrument that measures light frequencies emitted by lasers.

teledetection Information obtained from satellites.

toxic Poisonous.

INDEX

About the Author

Betty Burnett, author of several books for young people, lives in St. Louis, Missouri.

Photo Credits

Cover © Chuck St. John/Index Stock Imagery; pp. 11, 13, 34, 47, 49, 56, 58, 80, 97, 106, 107, 116, 125, 134 © AP Wide World; p. 15 © *Jacksonville Journal Courier*/The Image Works; pp. 20, 22 © Jonathan Blair/Corbis; p. 24 © Charles O'Rear/Corbis; pp. 25, 75, 77 © Joe Sohm/The Image Works; pp. 29, 31 © UNEP/Topham Picturepoint/The Image Works; pp. 38, 40 © Charles Shoffner/Index Stock Imagery; p. 43 © John Regan/Index Stock Imagery; p. 50 © Blair Howard/Index Stock Imagery; p. 59 © Josh Mitchell/Index Stock Imagery; p. 60 © Chris Minerva/Index Stock Imagery; pp. 66, 68 © Granata Press/The Image Works; p. 71 © Roger Ressmeyer/Corbis; pp. 84, 86 © Ed Lallo/Index Stock Imagery; p. 87 © Kevin Wilton/Corbis; pp. 93, 94 © Frank Pedrick/The Image Works; pp. 102, 104 © Peter Chapman/Index Stock Imagery; pp. 111, 113 © Bob Daemmrich/The Image Works; p. 114 © Matthew Borkoski/Index Stock Imagery; pp. 121, 123 © Bill Melton/Index Stock Imagery; pp. 130, 132 © Mike Greenlar/The Image Works.

Series Design

Evelyn Horovicz

Layout

Tahara Hasan

www.ingramcontent.com/pod-product-compliance
Lightning Source LLC
Chambersburg PA
CBHW050908210326
41597CB00002B/59

* 9 7 8 1 4 3 5 8 8 8 1 2 8 *